ERCOFTAC Series

Volume 32

ERCOFTAC (European Research Community on Flow, Turbulence and Combustion) was founded as an international association with scientific objectives in 1988. ERCOFTAC strongly promotes joint efforts of European research institutes and industries that are active in the field of flow, turbulence and combustion, in order to enhance the exchange of technical and scientific information on fundamental and applied research and design. Each year, ERCOFTAC organizes several meetings in the form of workshops, conferences and summer schools, where ERCOFTAC members and other researchers meet and exchange information. The *ERCOFTAC* series publishes the proceedings of ERCOFTAC meetings, which cover all aspects of fluid mechanics. The series comprises proceedings of conferences and workshops, and of textbooks presenting the material taught at summer schools. The series covers the entire domain of fluid mechanics, which includes physical modelling, computational fluid dynamics including grid generation and turbulence modelling, measuring techniques, flow visualization as applied to industrial flows, aerodynamics, combustion, geophysical and environmental flows, hydraulics, multi-phase flows, non-Newtonian flows, astrophysical flows, laminar, turbulent and transitional flows.

Indexed by SCOPUS, Google Scholar and SpringerLink.

Alexis Giauque · Christophe Corre · Martin White ·
Alberto Guardone
Editors

Proceedings of the 5th International Seminar on Non-Ideal Compressible Fluid Dynamics for Propulsion and Power

NICFD 2024
for Propulsion & Power

Springer

Editors
Alexis Giauque
Ecole Centrale de Lyon
Lyon, France

Christophe Corre
Ecole Centrale de Lyon
Lyon, France

Martin White
University of Sussex
Brighton, UK

Alberto Guardone
Politecnico di Milano
Milan, Italy

ISSN 1382-4309 ISSN 2215-1826 (electronic)
ERCOFTAC Series
ISBN 978-3-031-92694-5 ISBN 978-3-031-92695-2 (eBook)
https://doi.org/10.1007/978-3-031-92695-2

This Springer imprint is published by the registered company Springer Nature Switzerland AG
The registered company address is: Gewerbestrasse 11, 6330 Cham, Switzerland

Preface

The topic of non-ideal compressible fluid dynamics (NICFD) for propulsion and power deals with studying the fluid dynamics of compressible flows for which the ideal gas law does not apply, with an emphasis on reactive and non-reactive flows within power and propulsion systems. The study of NICFD effects finds application within industries that are critical to ensure the provision of clean and secure energy, which includes power generation, refrigeration, heat pumps and clean combustion amongst others. To understand the field of NICFD, we can break the topic into its constituent parts. The term non-ideal relates to the fundamental thermodynamic behaviour of unconventional fluids where the behaviour cannot be predicted assuming the fluid behaves as an ideal gas. Such thermodynamic behaviour can be observed in flows in supercritical states, flows close to the saturation curve, flows close to the critical point and two-phase vapour–liquid flows.

Secondly, we have compressible fluid dynamics which relates to the study of fundamental fluid dynamic aspects such as compressible high-speed flows, shock-wave formation, boundary layers, turbulence modelling, acoustics and phase change, amongst others. Combining this with non-ideal relates to understanding such fluid behaviour within thermodynamic regions where non-ideal effects are expected. Hand-in-hand with understanding these effects comes the need for computational fluid dynamic, experimental and measurement techniques, developed specifically for such flows.

Finally, power and propulsion relates to the application of this knowledge to practical engineering systems operating with non-ideal fluids. This relates to a range of technologies that are becoming, or are likely to become, increasingly important to ensure the provision of clean and secure energy, alongside efficient propulsion. This includes power cycles such as organic Rankine cycles and supercritical carbon dioxide power cycles, refrigeration systems, heat pumps, clean combustion, rocket engines, alongside subsequent system components such as turbomachinery, heat exchangers and combustors.

With this in mind, the seminar themes range from theoretical foundations, to advanced numerical and experimental practices, and to applications. The seminar provides an exciting platform to bring together researchers and scientists who are pioneering theoretical, numerical and experimental advancements in order to share, learn and discuss the latest insights in this field. The key themes of the conference include the following areas: experiments; fundamentals; numerical methods; optimisation and uncertainty quantification; critical and supercritical flows; turbulence and mixing; multi-component fluid flows; applications in ORC power systems; applications in supercritical CO_2 power systems; steam turbines; cryogenic flows; condensing flows in nozzles; cavitating flows and super- and trans-critical fluids in space propulsion.

The biannual NICFD seminar was established in 2016 due to the growing interest in NICFD effects, particularly stemming from advances in propulsion and power applications. The seminar aims to bring together researchers working in the field to discuss the latest advancements in the field. In 2019, the ERCOFTAC Special Interest Group on Non-Ideal Compressible Fluid Dynamics (SIG-49) was founded (SIG-49), which was setup to further promote the exchange of scientific information and to encourage and consolidate the interaction between NICFD researchers and professionals.

This volume contains the proceedings from NICFD2024: The 5th International Seminar on Non-Ideal Compressible Fluid Dynamics for Propulsion and Power, which was held during 30–31 October 2024, and hosted by the Laboratoire de Mecanique des Fluides et d'Acoustique at Ecole Centrale de Lyon, France. The published proceedings are composed of 19 papers reporting on the latest developments in the thematic areas of: Fundamentals; Numerical Modelling and Methods; Reacting Flows and Experiments.

The conference organisers are extremely grateful to everybody that helped make NICFD2024 a success. This includes all the authors of the submitted papers, as well as the reviewers and members of the scientific committee, who together ensured a high-quality of the contributions collected here, alongside the accompanying technical presentations. Thanks also go to the LMFA SAF and Centrale Innovation teams, the session chairs and the student helpers who ensured that the event ran smoothly.

Lyon, France — Alexis Giauque, Chair

Lyon, France — Christophe Corre
Brighton, UK — Martin White
Conference Co-chairs

Milan, Italy — Alberto Guardone, Founding Chair
January 2025

Contents

Fundamentals

Automatic Wave Curve Structure Detection for the Computation of Reference Solutions of Non-Classical Riemann Problems

Giuseppe Sirianni, Alberto Guardone, and Barbara Re

Abstract This paper outlines a solver for the computation of exact solutions of non-classical Riemann problems for the 1D Euler equations of gas-dynamics. The rarefaction and compression wave curve structures are automatically detected by minimizing two cost functions that embed all relevant physical constraints. The presented solver is automated but expensive, and is not meant to be used for the computation of fluxes in a Godunov-type solver, but rather for the computation of reference solutions for the validation of numerical tools. Some sample Riemann problems with non-classical waves are presented.

Keywords Non-classical gas dynamics · Non-convex thermodynamics · Exact Riemann solver · Exact solution · Wave curve structure

1 Introduction

The 1D Riemann problem of classical gas-dynamics has been studied extensively in the past century as it provides a building block for numerical tools aimed at simulating compressible fluid flows. It also provides a simple way to verify compressible CFD codes as the exact solution can be easily computed, also for generic equations of state, by studying the eigenstructure of the problem. After the seminal work of Bethe [1], Zeldovich [2] and Thompson [3] hinted at the possibility of a thermo-

G. Sirianni (✉) · A. Guardone · B. Re
Politecnico di Milano, Milan, Italy
e-mail: giuseppe.sirianni@polimi.it

A. Guardone
e-mail: alberto.guardone@polimi.it

B. Re
e-mail: barbara.re@polimi.it

A. Giauque et al. (eds.), *Proceedings of the 5th International Seminar on Non-Ideal Compressible Fluid Dynamics for Propulsion and Power*, ERCOFTAC Series 32,
https://doi.org/10.1007/978-3-031-92695-2_1

dynamic region where non-classical waves such as rarefaction shocks could arise, numerous works have been published outlining all possible non-classical wave curve configurations, see Smith [4], Cramer et al. [5] and Menikoff & Plohr [6]. One of the necessary conditions for non-classical waves to arise is that the wave curve must cross the thermodynamic region where the fundamental derivative of gas-dynamics [3] is negative, namely

$$\Gamma = 1 + \frac{\rho}{c}\left(\frac{\partial c}{\partial \rho}\right)_s < 0. \tag{1}$$

The construction of wave curves in the non-classical gas-dynamics regime has been thoroughly studied by Müller & Voss [7]. The implementation of the wave curve computation shown by Müller & Voss is complex due to the numerous implicit branching conditions that depend on the downstream and upstream states, but also on the wave curve path itself. In addition, their work extends to the non-smooth region of equations of state inside the saturation dome, which is outside of the scope of this paper. Fossati & Quartapelle [8] also computed non-classical solutions through a fourth-order polynomial approximation of the flux functions. In this work, we present an automated method to detect the rarefaction or compression wave curve structure through the minimization of a suitable cost function. The general Riemann problem solution method is described in Sect. 2. The generic wave curve structure is defined in Sect. 2.1 and the automated wave curve structure detection method is outlined in Sect. 2.2. The method is then used in Sect. 3 to compute exact non-classical solutions of 1D Riemann problems with fluids governed by the Span-Wagner EoS, implemented through the thermodynamic library CoolProp [9].

2 Methodology

The eigenstructure of the 1D unsteady Euler equation shows that the solution is always composed of three waves: a left- and a right-running acoustic wave, and a central contact discontinuity. This knowledge allows us to solve the Riemann problem by searching for the two star states $(\cdot)_{L,*}$ and $(\cdot)_{R,*}$ that satisfy the mechanical stability of the contact discontinuity (see Quartapelle et al. [10]), namely

$$\begin{cases} p_{L,*}(v_{L,*}; v_L, p_L) & = p_{R,*}(v_{R,*}; v_R, p_R) \\ u_{L,*}(v_{L,*}; u_L, v_L, p_L) & = u_{R,*}(v_{R,*}; u_R, v_R, p_R) \end{cases} \tag{2}$$

where v is the specific volume, p the pressure, u the velocity, $(\cdot)_L$ and $(\cdot)_R$ refer to the initial left and right states, and $(\cdot)_{L,*}$ and $(\cdot)_{R,*}$ to left and right star states adjacent to the contact discontinuity. For a cleaner notation, we use the semicolon to separate the variables of a function on the left, and its parameters on the right.

Starting from a generic initial state $k = L, R$, the solution along a wave curve can always be written in the following form

$$\begin{cases} p_{k,*}(v_{k,*};\, v_k,\, p_k) \\ u_{k,*}(v_{k,*};\, u_k,\, v_k,\, p_k) \end{cases} = \begin{cases} \begin{cases} p^{\text{rar}}(v_{k,*};\, v_k,\, p_k) \\ u^{\text{rar}}(v_{k,*};\, u_k,\, v_k,\, p_k) \end{cases} & \text{if} \quad v_{k,*} > v_k \\ \begin{cases} p^{\text{comp}}(v_{k,*};\, v_k,\, p_k) \\ u^{\text{comp}}(v_{k,*};\, u_k,\, v_k,\, p_k) \end{cases} & \text{if} \quad v_{k,*} < v_k \end{cases} \tag{3}$$

where $(\cdot)^{\text{rar}}$ refers to a rarefaction and $(\cdot)^{\text{comp}}$ to a compression.

2.1 Generic Wave Curve Structure Definition

Before defining the wave curves we need to define the solution along the basic building blocks that compose them, namely shock waves and isentropic fans. The solution across a shock-wave is given by the Rankine-Hugoniot locus which reads

$$RH(p,\, v;\, v_0,\, p_0) = e(v,\, p) - e(v_0,\, p_0) + \frac{1}{2}\,(p_0 - p)\,(v - v_0) = 0. \tag{4}$$

where e is the internal energy per unit mass. We can define the post-shock pressure $p^{RH}(v;\, v_0,\, p_0)$ as the solution of Eq. (4) for a given v, starting from a generic initial state $(\cdot)_0$. The post-shock velocity reads

$$u^{RH}(v;\, u_0,\, v_0,\, p_0) = u_0 \pm \sqrt{-(p^{RH}(v;\, v_0,\, p_0) - p_0)(v - v_0)}. \tag{5}$$

We can also define the shock speed from the Rankine-Hugoniot jump conditions as

$$\lambda^{RH}(v;\, u_0,\, v_0,\, p_0) = \frac{u^{RH}(v;\, u_0,\, v_0,\, p_0)/v - u_0/v_0}{1/v - 1/v_0}. \tag{6}$$

The pressure and velocity along an isentropic fan starting from a generic initial state $(\cdot)_0$ are computed as

$$\begin{cases} p^s(v;\, v_0,\, p_0) = p^s(v;\, s_0) = p(v,\, s_0) \\ u^s(v;\, u_0,\, v_0,\, p_0) = u^s(v;\, u_0,\, v_0,\, s_0) = u_0 \pm \displaystyle\int_{v_0}^{v} \frac{c(v',\, s_0)}{v'}\, dv' \end{cases} \tag{7}$$

where c is the speed of sound, s is the entropy per unit mass and $s_0 = s(v_0,\, p_0)$ is the initial entropy. The characteristic speeds along an isentropic fan are

$$\lambda^s(v;\, u_0,\, v_0,\, s_0) = u(v;\, u_0,\, v_0,\, s_0) \pm c(v,\, s_0). \tag{8}$$

Let us now define the generic structure of the rarefaction and compression wave curves. Firstly, in a generic rarefaction wave curve, we define two intermediate states v_1 and v_2 such that $v_0 \leq v_1 \leq v_2 \leq v_*$. A generic rarefaction wave curve can always be defined by piecing together an isentropic fan ($0 \to 1$), a Rankine-Hugoniot shock ($1 \to 2$), and another isentropic fan ($2 \to *$) in this exact order, see Fig. 1. The solution reads

$$\begin{cases} p^{\text{rar}}(v;\, v_0,\, p_0) \\ u^{\text{rar}}(v;\, u_0,\, v_0,\, p_0) \end{cases} = \begin{cases} \begin{cases} p^s(v;\, s_0) \\ u^s(v;\, u_0,\, v_0,\, s_0) \end{cases} & \text{if} \quad v \in [v_0,\, v_1) \\ \begin{cases} p^{RH}(v;\, v_1,\, p_1) \\ u^{RH}(v;\, u_1,\, v_1,\, p_1) \end{cases} & \text{if} \quad v \in [v_1,\, v_2] \;. \\ \begin{cases} p^s(v;\, s_2) \\ u^s(v;\, u_2,\, v_2,\, s_2) \end{cases} & \text{if} \quad v \in (v_2,\, v_*] \end{cases} \tag{9}$$

Conversely, in a generic compression wave curve, we define two intermediate states v_1 and v_2 such that $v_0 \geq v_1 \geq v_2 \geq v_*$. The compression wave curve can always be defined by piecing together a Rankine-Hugoniot shock ($0 \to 1$), an isentropic fan ($1 \to 2$), and another Rankine-Hugoniot shock ($2 \to *$) in this exact order, see Fig. 2. The solution reads

$$\begin{cases} p^{\text{comp}}(v;\, v_0,\, p_0) \\ u^{\text{comp}}(v;\, u_0,\, v_0,\, p_0) \end{cases} = \begin{cases} \begin{cases} p^{RH}(v;\, v_0,\, p_0) \\ u^{RH}(v;\, u_0,\, v_0,\, p_0) \end{cases} & \text{if} \quad v \in [v_0,\, v_1] \\ \begin{cases} p^s(v;\, s_1) \\ u^s(v;\, u_1,\, v_1,\, s_1) \end{cases} & \text{if} \quad v \in (v_1,\, v_2) \;. \\ \begin{cases} p^{RH}(v;\, v_2,\, p_2) \\ u^{RH}(v;\, u_2,\, v_2,\, p_2) \end{cases} & \text{if} \quad v \in [v_2,\, v_*] \end{cases} \tag{10}$$

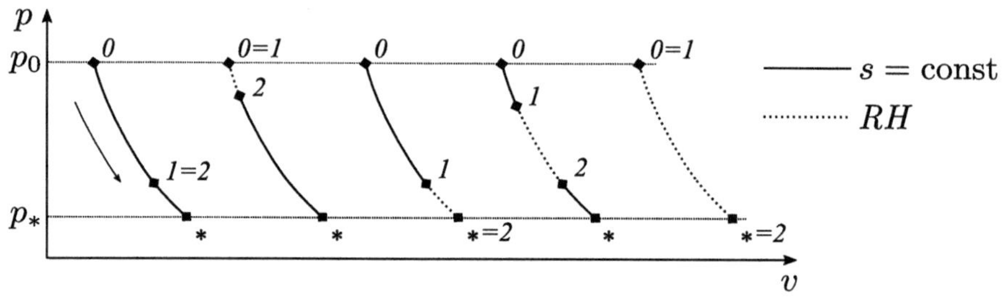

Fig. 1 All possible rarefaction wave curve configurations. From left to right, classical fan, shock-fan, fan-shock, fan-shock-fan, shock

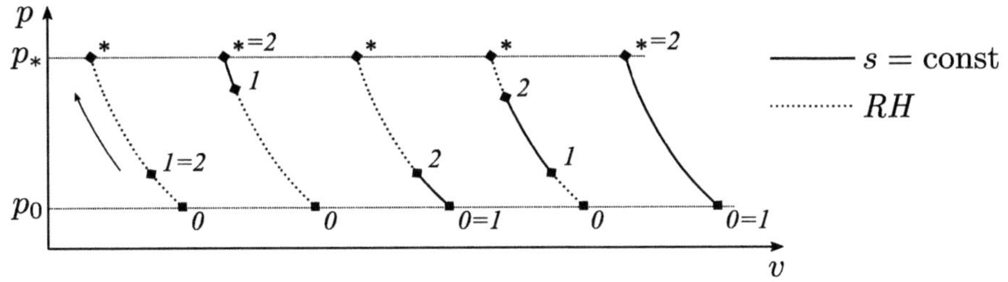

Fig. 2 All possible compression wave curve configurations. From left to right, classical shock, shock-fan, fan-shock, shock-fan-shock, fan

2.2 *Wave Curve Structure Detection*

We propose here an automated method to detect the correct wave curve from given upstream conditions and a certain downstream specific volume. To do so one first needs to define a cost function for each wave curve, such that they are equal to zero if and only if all physical constraints are met, namely $F^{\text{comp}}(v_1, v_2; v_*, u_0, v_0, p_0) = 0$ and $F^{\text{rar}}(v_1, v_2; v_*, u_0, v_0, p_0) = 0$. The constraints for a rarefaction wave curve are:

1. The $(0 \to 1)$ and $(2 \to *)$ fan's characteristic speeds $\lambda^s(v; u_0, v_0, s_0)$ for $v \in [v_0, v_1)$, and $\lambda^s(v; u_2, v_2, s_2)$ for $v \in (v_2, v_*]$ must be monotonic.
2. If the $(1 \to 2)$ shock is present, the entropy jump $(s_2 - s_1)$ must be positive.
3. If the $(1 \to 2)$ shock is present, all characteristic speeds left and right of the shock must go into the shock.
4. If the $(0 \to 1 \to 2)$ fan-shock is present, the $(1 \to 2)$ shock and the $(0 \to 1)$ fan's last characteristic must have equal speed, namely $\lambda^{RH}(v_2; u_1, v_1, p_1) = \lambda^s(v_1; u_0, v_0, s_0)$.
5. If the $(1 \to 2 \to *)$ shock-fan is present, the $(1 \to 2)$ shock and the $(2 \to *)$ fan's first characteristic must have equal speed, namely $\lambda^{RH}(v_2; u_1, v_1, p_1) = \lambda^s(v_2; u_2, v_2, s_2)$.

The constraints for a compression wave curve are:

1. If the $(1 \to 2)$ fan is not present, a single $(0 \to *)$ compression shock is present and the entropy jump $(s_* - s_0)$ must be positive.
2. If the $(0 \to 1 \to 2)$ shock-fan is present, the entropy jump $(s_1 - s_0)$ must be positive.
3. If the $(1 \to 2 \to *)$ fan-shock is present, the entropy jump $(s_* - s_2)$ must be positive.
4. If the $(1 \to 2)$ fan is present, characteristic speeds $\lambda^s(v; u_1, v_1, s_1)$ for $v \in (v_1, v_2)$ must be monotonic.
5. If the $(0 \to 1 \to 2)$ shock-fan is present, the $(0 \to 1)$ shock and the $(1 \to 2)$ fan's first characteristic must have equal speed, $\lambda^{RH}(v_1; u_0, v_0, p_0) = \lambda^s(v_1; u_1, v_1, s_1)$.

6. If the $(1 \to 2 \to *)$ fan-shock is present, the $(2 \to *)$ shock and the $(1 \to 2)$ fan's last characteristic must have equal speed, $\lambda^{RH}(v_*; u_2, v_2, p_2) = \lambda^s(v_2; u_1, v_1, s_1)$.
7. If the $(0 \to 1)$ shock is present, all characteristic speeds left and right of the shock must go into the shock.
8. If the $(2 \to *)$ shock is present, all characteristic speeds left and right of the shock must go into the shock.

We define the cost function for the left and right wave curves as

$$F_k(v_{k,1}, v_{k,2}; v_{k,*}, u_k, v_k, p_k) = \begin{cases} F^{\text{rar}}(v_{k,1}, v_{k,2}; v_{k,*}, u_k, v_k, p_k) & \text{if} \quad v_{k,*} > v_k \\ F^{\text{comp}}(v_{k,1}, v_{k,2}; v_{k,*}, u_k, v_k, p_k) & \text{if} \quad v_{k,*} < v_k \end{cases}. \tag{11}$$

If one obtains the roots $(v_{L,1}, v_{L,2})$ and $(v_{R,1}, v_{R,2})$ of the cost functions, the solution along the left and right wave curves are defined by Eqs. (9), (10) and (3). This can be therefore used at each iteration of the solution procedure for the contact discontinuity system Eq. (2) to find the solution of the Riemann problem. Note that one needs to define a certain threshold ϵ below which a component of the wave curve is considered nonexistent and therefore skipped. For example, if for a rarefaction wave one finds that $(v_2 - v_1) < \epsilon$, then the rarefaction shock portion of the wave curve $(1 \to 2)$ is considered non-existent (or not detected). Since we need to perform numerical integral along isentropic fans to solve Eq. (7), it is convenient to use the same step size dv used for integration for the threshold ϵ, which in our tests is $dv = \epsilon = 10^{-5}$ m^3/kg.

3 Results

We use the automatic wave curve detection presented in this paper to solve two non-classical Riemann problems that we showcase here. We use a tolerance of 10^{-15} for the root-finding algorithm used in solving the contact-discontinuity system Eq. (2) and the cost functions of the wave curve detection method Eq. (11). For a rough measure of the computational cost, the solution of each test using the Span-Wagner EoS through the CoolProp [9] thermodynamic library took around 55–90 s on a single core at 4.8 GHz.

The first test is an MD_4M Riemann problem with the initial data given in Table 1a. The solution plotted in Fig. 3 is composed of two rarefaction waves, but while the left wave doesn't cross the $\Gamma < 0$ region and is therefore a classical fan, the right wave curve completely crosses the $\Gamma < 0$ region and the solution degenerates into a non-classical fan-shock-fan. We can see, in the pressure-density thermodynamic plot in Fig. 3, that the pre- and post-shock states of the non-classical shock lie on the double sonic shock locus (DSL) as defined in Zamfirescu et al. [11]. This confirms that the proposed wave curve detection procedure is indeed able to correctly capture the

Table 1 Riemann problems initial conditions

(a) MD_4M

	$x < 0.5$ m	$x > 0.5$ m
ρ	200 kg/m^3	335 kg/m^3
u	−70 m/s	90 m/s
P	13 bar	10 bar

(b) D_5

	$x < 0.5$ m	$x > 0.5$ m
ρ	137 kg/m^3	190.4 kg/m^3
u	0 m/s	0 m/s
P	9.25 bar	10.28 bar

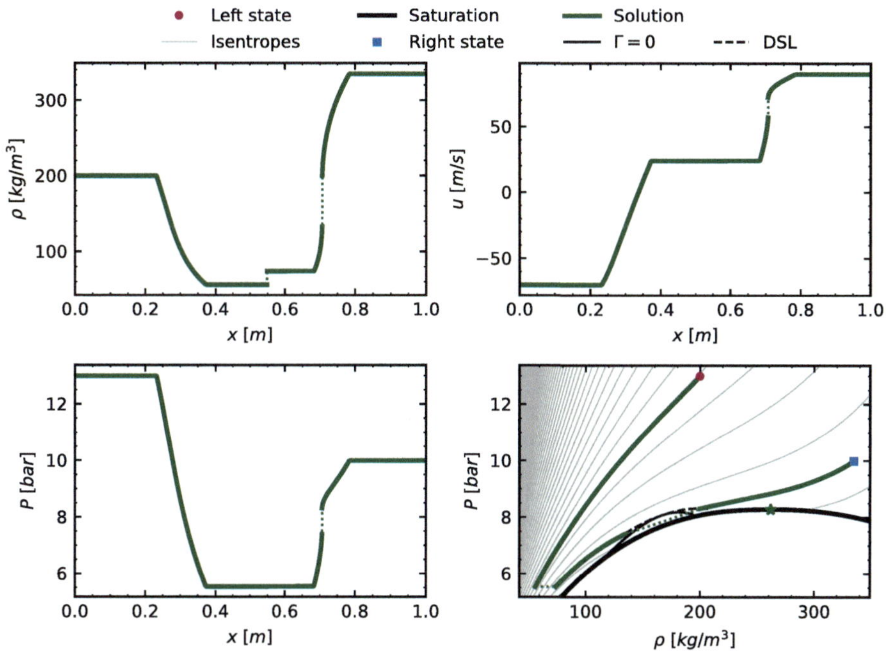

Fig. 3 MD_4M Riemann problem solution composed of a classical left running rarefaction fan, a central contact discontinuity, and non-classical right running rarefaction fan-shock-fan. Plots of the density, velocity and pressure profiles at $t = 0.002$ s, and path on the density-pressure thermodynamic plane, with the double sonic shock locus (DSL) as defined by Zamfirescu et al. [11]

non-classical double sonic rarefaction shock embedded between the two rarefaction fans.

The second test is a D_5 Riemann problem with the initial data given in Table 1b. The solution plotted in Fig. 4 is composed of a left-running compression wave and a right running rarefaction wave. Since the initial states are close to the $\Gamma < 0$ region, the thermodynamic states at the left and right of the contact discontinuity lie within it. The resulting left-running compression wave is a non-classical shock-fan, where the initial compression shock degenerates into a compression fan only as the shock speed matches the speed of the first fan's characteristic, which occurs at a thermodynamic point well within the $\Gamma < 0$ region. Conversely, the right running rarefaction wave

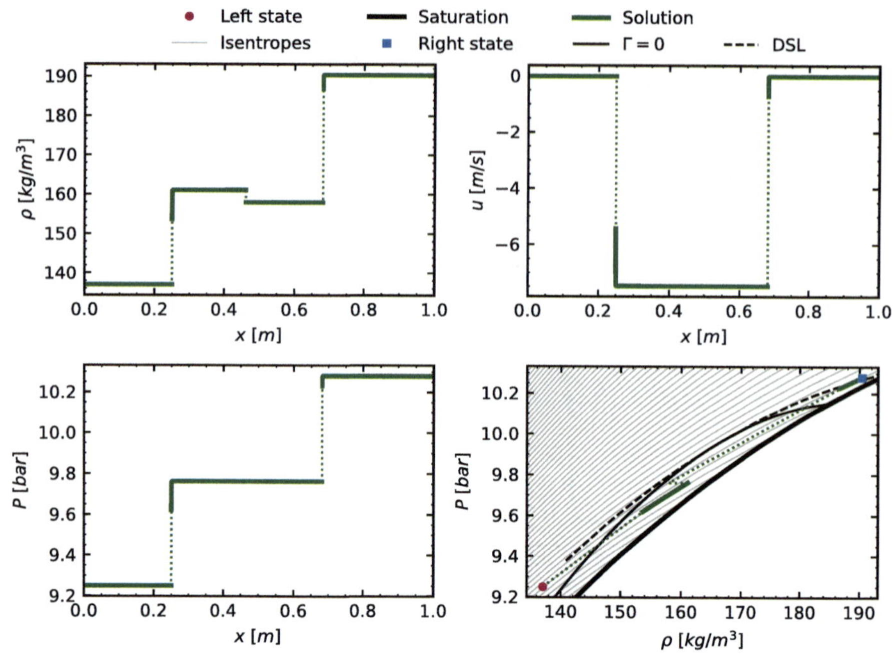

Fig. 4 D_5 Riemann problem solution composed of a non-classical left running compression shock-fan, a central contact discontinuity, and a non-classical right running rarefaction fan-shock. Plots of the density, velocity, and pressure profiles at $t = 0.005$ s, and path on the density-pressure thermodynamic plane, with the double sonic shock locus (DSL) as defined by Zamfirescu et al. [11]

is a non-classical fan-shock. Analogously to the left-running compression wave, the initial fan degenerates into a compression shock only when the last fan's characteristic speed matches the compression shock, which this time occurs well outside the $\Gamma < 0$ region.

4 Conclusions and Future Works

We presented an automated method for detecting the correct wave curve structure in the non-classical gas-dynamics regime and applied it to the solution of two 1D Riemann problems using the Span-Wagner EoS through the CoolProp thermodynamic library. The method is valid for a generic equation of state but is also expensive, and therefore it is unsuitable for a Godunov-type solver. Future work will be focusing on the study of Riemann problems of equilibrium mixtures at varying compositions, intending to perform verification of multi-phase models not based on a mixture equation of state but relaxation terms that drive the phases towards equilibrium.

Acknowledgements B. Re acknowledges the financial support received by the European Union under the NextGenerationEU action, implemented through the PRIN project NICTURB–CUP: D53D23003290001. G. Sirianni acknowledges the financial support of the European Union under the NextGenerationEU action.

References

1. Bethe, H.A.: On the theory of shock waves for an arbitrary equation of state. In: Classic Papers in Shock Compression Science, pp. 421–495. Springer, New York (1942)
2. Zeldovich, Y.B.: On the possibility of rarefaction shock waves. In: Selected Works of Yakov Borisovich Zeldovich, Volume I, pp. 152–154. Princeton University Press (1946)
3. Thompson, P.A.: A fundamental derivative in gasdynamics. Phys. Fluids **14**(9), 1843 (1971)
4. Smith, R.G.: The riemann problem in gas dynamics. Trans. Am. Math. Soc. **249**(1), 1–50 (1979)
5. Cramer, M.S., Sen, R.: Shock formation in fluids having embedded regions of negative nonlinearity. Phys. Fluids **29**(7), 2181–2191 (1986)
6. Menikoff, R., Plohr, B.J.: The riemann problem for fluid flow of real materials. Rev. Mod. Phys. **61**, 75–130 (1989)
7. Müller, S., Voss, A.: The riemann problem for the euler equations with nonconvex and nonsmooth equation of state: Construction of wave curves. SIAM J. Sci. Comput. **28**(2), 651–681 (2006)
8. Fossati, M., Quartapelle, L.: The riemann problem for hyperbolic equations under a nonconvex flux with two inflection points (2014)
9. Bell, I.H., Wronski, J., Quoilin, S., Lemort, V.: Pure and pseudo-pure fluid thermophysical property evaluation and the open-source thermophysical property library coolprop. Ind. Eng. Chem. Res. **53**(6), 2498–2508 (2014)
10. Quartapelle, L., Castelletti, L., Guardone, A., Quaranta, G.: Solution of the riemann problem of classical gasdynamics. J. Comput. Phys. **190**(1), 118–140 (2003)
11. Zamfirescu, C., Guardone, A., Colonna, P.: Admissibility region for rarefaction shock waves in dense gases. J. Fluid Mech. **599**, 363–381 (2008)

Conservative Temperature Update for the Efficient Simulation of Single and Multi-Phase Flows using the Span-Wagner Equation of State

Giuseppe Sirianni, Alberto Guardone, Barbara Re, and Rémi Abgrall

Abstract In this paper we present a temperature update scheme that conserves the total energy and is valid for an arbitrary equation of state and spatial discretization. We derive the scheme for the single-phase Euler equations and present results for a shock tube spanning from super-critical to near-saturation conditions, measuring total energy imbalance and computational costs associated with thermodynamic evaluations. We also showcase non-classical results obtained with a multi-phase extension of the scheme for the full disequilibrium Baer-Nunziato equations.

Keywords Temperature update · Conservative scheme · Span-Wagner equation of state · Multi-phase flows · Non-ideal compressible fluid dynamics (NICFD) · Non-classical Riemann problem

1 Introduction

Non-ideal compressible fluid dynamics (NICFD) is driven by the fluid's complex thermodynamic behavior, which calls for accurate modeling of its thermodynamic properties. For this reason, equations of state (EoS) such as the Span-Wagner EoS [1] have been widely adopted. The main drawback when employing this EoS is the computational cost because it is naturally expressed as a function of density and temperature (ρ, T) whereas, the thermodynamic variables directly available from

G. Sirianni (✉) · A. Guardone · B. Re
Politecnico di Milano, Milan, Italy
e-mail: giuseppe.sirianni@polimi.it

A. Guardone
e-mail: alberto.guardone@polimi.it

B. Re
e-mail: barbara.re@polimi.it

R. Abgrall
Universität Zürich, Zürich, Switzerland
e-mail: remi.abgrall@math.uzh.ch

A. Giauque et al. (eds.), *Proceedings of the 5th International Seminar on Non-Ideal Compressible Fluid Dynamics for Propulsion and Power*, ERCOFTAC Series 32,
https://doi.org/10.1007/978-3-031-92695-2_2

the fluid solver are density and energy (ρ, E). A common strategy used to mitigate computational costs of the EoS in NICFD simulations has been the tabulation of thermodynamic properties in so-called lookup tables (LUT), see Pini et al. [2]. LUTs have been great tools that have opened the possibility for large-scale computational fluid dynamics (CFD) simulations in the NICFD regime with accurate thermodynamic models that would have otherwise been unfeasible, such as the shape optimization of 3D axial turbine stages. Another strategy employed to mitigate the high computational cost associated with the evaluation of the thermodynamic properties using complex EoS is solving the governing equations in primitive form, to avoid the solution of the non-linear equation needed to retrieve the thermodynamic state from the total energy. A large body of work has been dedicated to primitive formulations of the Euler equations, e.g., [3, 4]. The main issue with primitive formulations is the loss of conservation, which is a foundational property that any scheme applied to the compressible regime should possess. A non-conservative scheme may converge to a non-entropic solution if discontinuities are present, capturing incorrect shock speeds and jump conditions.

Recently, Abgrall [5] introduced a residual correction approach that recovers conservation if the internal energy E has a linear dependency with respect to the thermodynamic variable φ that is updated instead of total energy E^t, therefore if $(\partial E/\partial \varphi)_\rho = E_{\varphi,\rho} = \text{const}$. Inspired by the work of Abgrall, a total energy conservative update scheme for a generic primitive variable φ and valid for arbitrary equation of state and spatial discretization has been presented in Sirianni et al. [6]. The present paper describes the scheme that updates the temperature ($\varphi = T$), which is the most promising choice for NICFD applications. Section 2 briefly shows the derivation of the scheme for the single-phase Euler equations. In Sect. 3, we first show the results of a single-phase shock-tube, focusing particularly on total energy imbalance and computational costs, and then we showcase a multi-phase extension of the scheme tested on a two-phase non-classical Riemann problem. In Sect. 4, we draw some brief conclusions and outline future developments.

2 Numerical Scheme

The conservative form of the Euler equations of gas-dynamics reads

$$\frac{\partial}{\partial t}\underbrace{\begin{bmatrix}\rho \\ \rho\mathbf{u} \\ E^t\end{bmatrix}}_{\mathbf{q}} + \nabla\cdot\underbrace{\begin{bmatrix}\rho\mathbf{u} \\ (\rho\mathbf{u}\otimes\mathbf{u} + P\boldsymbol{I}) \\ \left(E^t + P\right)\mathbf{u}\end{bmatrix}}_{\boldsymbol{F}} = \mathbf{0}\,, \tag{1}$$

where ρ is the density, $\mathbf{u}$ is the velocity, P is the pressure and $E^t = E + \rho\mathbf{u}^2/2$ is the total energy. We also denote with $\mathbf{q}$ the conservative variables and $\boldsymbol{F}$ the fluxes.

In this paper, we consider a finite-volume solver with MUSCL reconstruction using the Barth-Jespersen [7] slope limiter. In time, we use an explicit Euler update scheme, that for a generic variable q in a generic control volume C_i reads

$$\Phi_i^q = \sum_{C_j \in \partial C_i} A_{i,j} F^q \left(\mathbf{q}_i, \mathbf{q}_j\right) \qquad \rightarrow \qquad q_i^{n+1} = q_i^n - \frac{\Delta t}{C_i} \Phi_i^q \,, \tag{2}$$

where Φ is the spatial residual, C_j the control volume neighbouring C_i, and $A_{i,j}$ their shared area. We use Eq. (2) to directly update density $q_i = \rho_i$ and momentum $\mathbf{q}_i = \rho \mathbf{u}_i$, while for the total energy we only compute $\Phi_i^{E^t}$ and use a different scheme for the update, which we present in the next paragraph. Note that the spatial discretization is completely embedded in the definition of the spatial residual Φ, and, as long as it can be written in the generic form given in Eq. (2), the choice of the discretization technique does not affect the scheme presented in this work.

For a cleaner notation, we now drop subscript $(\cdot)_i$ as we focus on a single control volume. The total energy and its time derivative can be written as

$$E^t = E(\rho, T) + \frac{1}{2}\rho \mathbf{u}^2 \qquad \rightarrow \qquad \frac{\partial E^t}{\partial t} = E_{T,\rho}\frac{\partial T}{\partial t} + E_{\rho,T}\frac{\partial \rho}{\partial t} + \frac{1}{2}\frac{\partial \rho \mathbf{u}^2}{\partial t}\,, \tag{3}$$

where in general, $E_{T,\rho} = (\partial E/\partial T)_\rho$ and $E_{\rho,T} = (\partial E/\partial \rho)_T$ are functions of time. By integrating Eq. (3) between time t^n and t^{n+1}, we can obtain an approximate expression for the jump ΔE^t assuming the derivatives of internal energy as constants, with values $\overline{E_{T,\rho}}$ and $\overline{E_{\rho,T}}$

$$\begin{aligned} \int_{t^n}^{t^{n+1}} \frac{\partial E^t}{\partial t} dt &= \int_{t^n}^{t^{n+1}} \left[E_{T,\rho}\frac{\partial T}{\partial t} + E_{\rho,T}\frac{\partial \rho}{\partial t} + \frac{1}{2}\frac{\partial \rho \mathbf{u}^2}{\partial t}\right] dt \\ \Delta E^t &\simeq \underbrace{\overline{E_{T,\rho}}\Delta T + \overline{E_{\rho,T}}\Delta \rho}_{\Delta E^{\text{approx}}} + \frac{1}{2}\Delta \rho \mathbf{u}^2 \,. \end{aligned} \tag{4}$$

We can plug this expression in the total energy conservation law integrated over the control volume C, thus obtaining

$$\int_C \Delta E^t dV + \Delta t \Phi^{E^t} \simeq \int_C \left(\overline{E_{T,\rho}}\Delta T + \overline{E_{\rho,T}}\Delta \rho + \frac{1}{2}\Delta \rho \mathbf{u}^2 \right) dV + \Delta t \Phi^{E^t} = 0\,. \tag{5}$$

Then, we update the temperature using the same explicit formula Eq. (2) used for density and momentum:

$$T^{n+1} = T^n - \frac{\Delta t}{C}\Phi^T \quad \rightarrow \quad \Delta T = -\frac{\Delta t}{C}\Phi^T \,, \tag{6}$$

where the pseudo-spatial residual for temperature Φ^T is unknown. By substituting the temperature jump ΔT from Eq. (6) into the approximate total energy conservation

Table 1 Initial conditions for the Riemann problems using the Span-Wagner EoS through CoolProp [8]

(a) CO_2 *classical single-phase test*

	$x < 0.5$ m	$x > 0.5$ m
ρ	350 kg/m^3	100 kg/m^3
u	0 m/s	0 m/s
P	12 MPa	4 MPa

(b) MD_4M *non-classical two-phase test*

	$x < 0.2$ m	$x > 0.2$ m
Fluid	MD$_4$M vapour	Toluene vapour($C_6H_5CH_3$)
ρ	340 kg/m^3	8 kg/m^3
u	0 m/s	50 m/s
P	1 MPa	0.345 MPa

Eq. (5) and inverting we obtain an expression for Φ^T:

$$\Phi^T(\overline{E_{T,\rho}}, \overline{E_{\rho,T}}) = \frac{C}{\Delta t \int_C \overline{E_{T,\rho}} dV} \left[\int_C \left(\overline{E_{\rho,T}} \Delta\rho + \frac{1}{2} \Delta\rho \mathbf{u}^2 \right) dV + \Delta t \Phi^{E^t} \right]. \tag{7}$$

Using Eqs. (6) and (7) directly would not result in total energy conservation due to the linearization of $\overline{E_{T,\rho}}$ and $\overline{E_{\rho,T}}$. To retrieve conservation, we start by defining a linearised thermodynamic state $(\overline{\rho}, \overline{T})$, and since the density at the next time step t^{n+1} is already known from Eq. (2), we fix the value $\overline{\rho} = \left(\rho^{n+1} + \rho^n\right)/2$. We impose thermodynamic consistency on the internal energy derivatives by computing them at the linearization state, therefore $\overline{E_{T,\rho}} = E_{T,\rho}(\overline{\rho}, \overline{T})$ and $\overline{E_{\rho,T}} = E_{\rho,T}(\overline{\rho}, \overline{T})$. The only unknown is now the linearization temperature $\overline{T}$. We recall the approximate internal energy jump ΔE^{approx} from Eq. (4) and define its thermodynamic counterpart ΔE^{tmd} as

$$\begin{cases} \Delta E^{\text{approx}}(\overline{T}) = E_{T,\rho}(\overline{\rho}, \overline{T}) \Delta T(\overline{T}) + E_{\rho,T}(\overline{\rho}, \overline{T}) \Delta\rho \\ \Delta E^{\text{tmd}}(\overline{T}) = E(\rho^{n+1}, T^{n+1}(\overline{T})) - E(\rho^n, T^n) \end{cases} \rightarrow \quad F(\overline{T}) = \frac{\Delta E^{\text{tmd}}(\overline{T}) - \Delta E^{\text{approx}}(\overline{T})}{E(\rho^n, T^n)}, \tag{8}$$

where $F(\overline{T})$ is a function whose root identifies the linearization temperature $\overline{T}$ that provides total energy conservation. To find the value of $\overline{T}$ that satisfies Eq. (8) we use the secant root search algorithm with tolerance $|F| \leq 10^{-14}$, and as initial bounds $\left[T^n, \tilde{T}^{n+1}\right]$ with $\tilde{T}^{n+1}$ computed from Eqs. (6) and (7) using $\overline{T} = T^n$. The final value of $\overline{T}$ is then used to update T^n to T^{n+1} using Eqs. (6) and (7).

3 Results

We show results obtained using the presented scheme with the Span-Wagner EoS [1] implemented through the CoolProp thermodynamic library [8]. The first test is a single-phase CO_2 classical shock tube, with initial data given in Table 1a and solved on an equispaced mesh with $n_x = 400$ points and at $t_{\text{final}} = 1$ ms. We use the HLLC approximate Riemann solver [9] with the Vinokur-Montagné Roe average [10]. The solution plotted in Figs. 1 and 2 spans from supercritical to conditions near saturation. Results overlap everywhere when comparing the presented temperature update (T Update) and the standard total energy update, both evaluating the EoS analytically (E^t Update) and through LUTs ($E^{t,(\mathrm{LUT})}$ Update). The dimensionless imbalance of total energy $I^{E^t}(t)$ in the domain plotted in Fig. 2 is close to machine precision for all update schemes, with the E^t update being closer to zero by construction.

Computational costs associated with thermodynamic evaluations are reported in Fig. 3. The costs are measured and subdivided into three main categories: costs

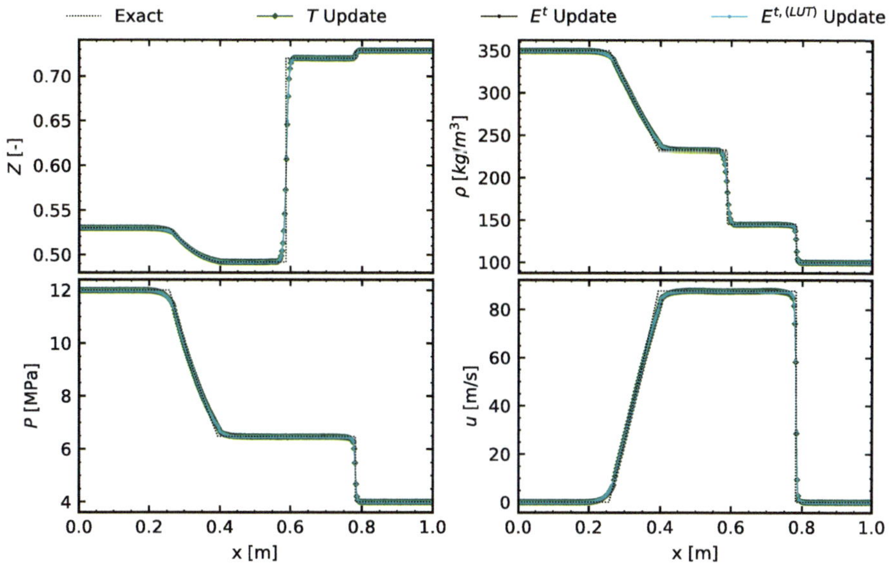

Fig. 1 Results at $t = 1$ ms for the CO_2 shock-tube test using the Span-Wagner EoS with $n_x = 400$. Spatial profiles of the compressibility factor Z, density ρ, pressure P and velocity u

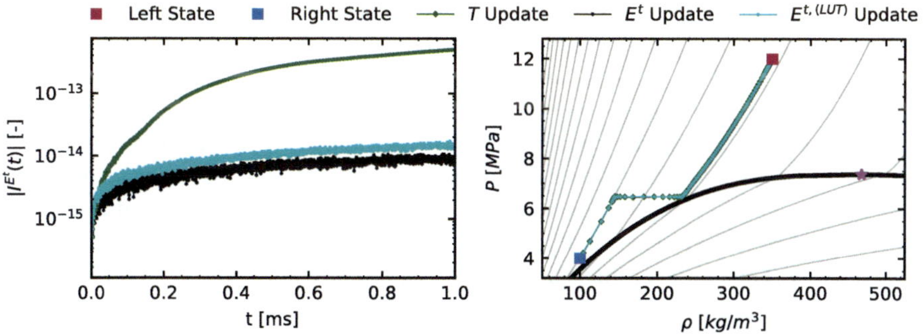

Fig. 2 Results at $t = 1$ ms for the CO_2 shock-tube test using the Span-Wagner EoS with $n_x = 400$. Left: time evolution of the absolute value of the total energy imbalance; right: solution on the density-pressure thermodynamic plane (ρ, P) with isentropes in grey and critical point in pink

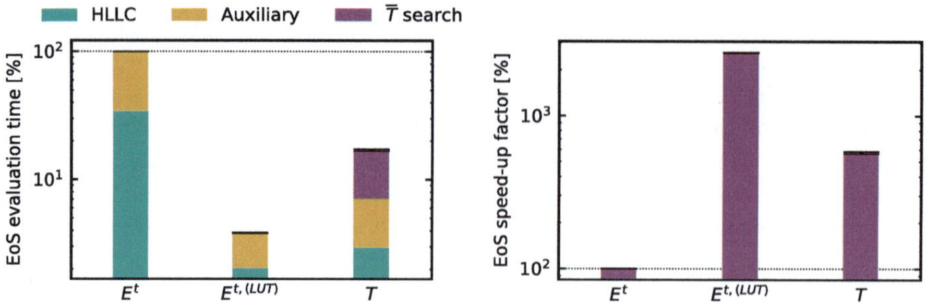

Fig. 3 Computational costs of thermodynamic evaluations for the CO2 shock-tube test using the Span-Wagner EoS

associated with the HLLC approximate Riemann solver and upwind residual computations, auxiliary calls (eg. CFL time step computation), and calls required by the $\overline{T}$ search introduced in this paper. The measured speed-up is around 700% for the proposed T update compared to the standard E^t update, while compared to the use of LUTs it is not as advantageous, mainly due to the additional EoS calls introduced in the $\overline{T}$ search. The maximum number of iterations used for the $\overline{T}$ search is $n_{iter} = 5$, while the average across all cells and time steps is $n_{iter} = 1.00473$.

The second test presented is a two-phase MD_4M-toluene non-classical Riemann problem, with initial data given in Table 1b and solved on an equispaced $n_x = 1000$ mesh with $t_{\text{final}} = 2.5$ ms. We use an extension of the presented T update scheme for the full-disequilibrium two-phase model of Baer-Nunziato [11] which reads

$$\frac{\partial}{\partial t}\begin{bmatrix}\alpha_1\\ \alpha\rho_k\\ \alpha\rho\mathbf{u}_k\\ \alpha E_k^t\end{bmatrix} + \nabla\cdot\begin{bmatrix}\mathbf{0}\\ \alpha\rho\mathbf{u}_k\\ \alpha_k\left(\rho\mathbf{u}_k\otimes\mathbf{u}_k + P_k\boldsymbol{I}\right)\\ \alpha_k\left({E^t}_k + P_k\right)\mathbf{u}_k\end{bmatrix} + \begin{bmatrix}\mathbf{u}_I\\ \mathbf{0}\\ (-1)^k P_I\boldsymbol{I}\\ (-1)^k P_I\mathbf{u}_I\end{bmatrix}\cdot\nabla\alpha_1 = \begin{bmatrix}\mu\Delta_1 P\\ 0\\ -\lambda\Delta_k\mathbf{u}\\ -\lambda\mathbf{u}_I\Delta_k\mathbf{u} - \mu P_I\Delta_k P\end{bmatrix}. \tag{9}$$

with $k = 1, 2$ the phase. This model is composed by an advection equation for the volume fraction α and two sets of evolution equations for the partial densities, momentum, and total energies. Compared to the single-phase Euler equations, these contain additional non-conservative phase coupling terms that depend on the interface velocity $\mathbf{u}_I$ and pressure P_I which require additional closure. The relaxation process towards mechanical equilibrium is modeled through source terms containing the velocity and pressure difference between phases, eg. $\Delta_2\mathbf{u} = (\mathbf{u}_2 - \mathbf{u}_1)$ and $\Delta_1 P = (P_1 - P_2)$. The relaxation speed, which we will assume infinitely fast, is governed by the velocity and pressure relaxation constants, $\lambda = \infty$ and $\mu = \infty$, respectively. The model embeds mixture conservation without the need for a description of the mixture EoS [12]. Each phase is governed by its own EoS, and by summing the equations over the phases $k = 1, 2$ we obtain the conservative Euler equations for the mixture. To solve the equations we perform a Strang splitting of the hyperbolic operator and the relaxation ODEs which are, respectively, the terms left and right of the equality in Eq. (9). We first solve the former using an extension of the

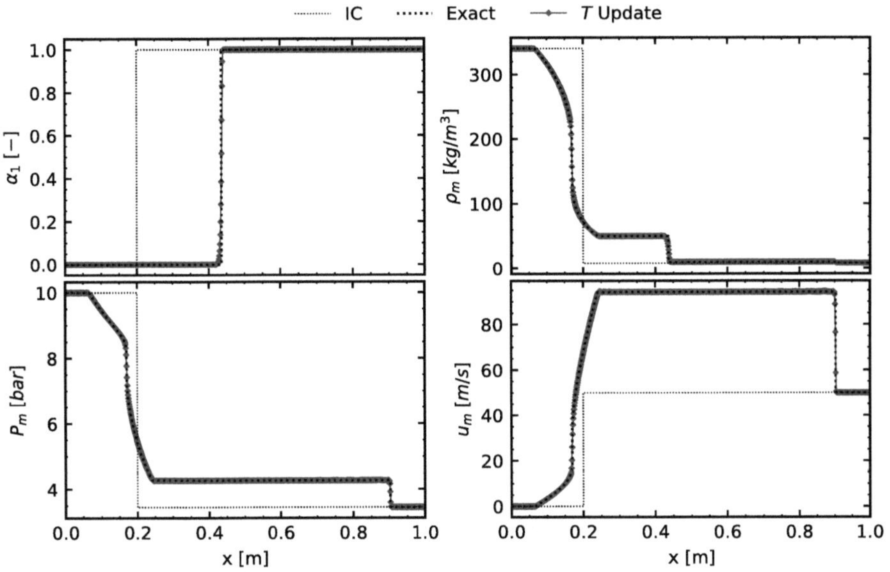

Fig. 4 Results at $t = 2.5$ ms for the MD_4M-toluene two-phase non-classical shock-tube test using the Span-Wagner EoS with $n_x = 1000$. Spatial profiles of the toluene volume fraction α_1, mixture density $\rho_m = (\alpha\rho_1 + \alpha\rho_2)$, mixture pressure $P_m = (\alpha_1 P_1 + \alpha_2 P_2)$ and mixture velocity $u_m = (\alpha\rho\mathbf{u}_1 + \alpha\rho\mathbf{u}_2)/\rho_m$

scheme presented in this paper, with the HLLC-type solver of Lochon et al. [13] and the Vinokur-Montagné Roe average [10]. The relaxation ODEs are subsequently integrated using the approach of Chiocchetti and Muller [14]. The solution is plotted in space in Fig. 4 and on the thermodynamic plane in Fig. 5. We can see a right-running classical compression shock in toluene, and a left-traveling rarefaction wave in MD_4M crossing the $\Gamma < 0$ region, resulting in a non-classical fan-shock-fan composite wave. Also, the contact discontinuity corresponds with the material contact between the two fluids. All waves are well captured with minimal numerical dissipation.

4 Conclusions and Future Works

We presented a temperature update scheme that is total energy conservative and valid for a generic equation of state and spatial discretization. Solutions of a CO_2 shock-tube using the Span-Wagner EoS show that total energy is conserved up to machine precision and the solution for the presented scheme overlaps the standard total energy update, with a measured computational speed-up of up to 700%. We also presented results for a two-phase non-classical Riemann problem using an extension of the presented temperature update scheme to a two-phase disequilibrium model. All

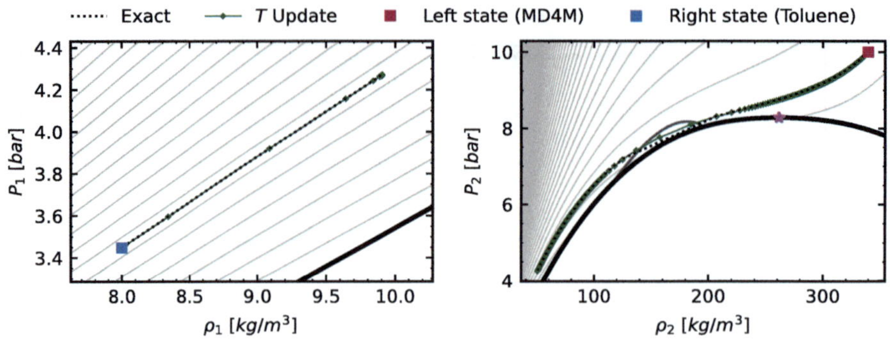

Fig. 5 Results at $t = 2.5$ ms for the MD_4M-toluene two-phase non-classical shock-tube test using the Span-Wagner EoS with $n_x = 1000$. Solution on the density-pressure thermodynamic planes for toluene (ρ_1, P_1) and MD_4M (ρ_2, P_2) with isentropes in grey and critical point in pink

results obtained agree with analytical solutions. Future work will focus on applying the presented scheme to industrial applications such as ORC turbines and adding a temperature-based finite rate phase-transition solver to the multi-phase model.

Acknowledgements B. Re acknowledges the financial support received by the European Union's Horizon Europe programme under the MSCA-2021-PF grant agreement Id 101066019 (project NI2PhORC). G. Sirianni acknowledges the financial support of the European Union under the NextGenerationEU action.

References

1. Span, R., Wagner, W.: A new equation of state for carbon dioxide covering the fluid region from the triple-point temperature to 1100 k at pressures up to 800 MPa. J. Phys. Chem. Ref. Data **25**(6), 1509–1596 (1996)
2. Pini, M., Spinelli, A., Persico, G., Rebay, S.: Consistent look-up table interpolation method for real-gas flow simulations. Comput. Fluids **107**, 178–188 (2015)
3. Toro, E.F., Millington, R.C., Nejad, L.A.M.: Primitive upwind numerical methods for hyperbolic partial differential equations. In: Charles-Henri, B. (ed.) Sixteenth International Conference on Numerical Methods in Fluid Dynamics, pp. 421–426. Springer, Berlin, Heidelberg (1998)
4. Bijl, H., Wesseling, P.: A unified method for computing incompressible and compressible flows in boundary-fitted coordinates. J. Comput. Phys. **141**(2), 153–173 (1998)
5. Abgrall, R.: Staggered schemes for compressible flow: A general construction. SIAM J. Sci. Comput. **46**(1), A399–A428 (2024)
6. Sirianni, G., Guardone, A., Re, B., Abgrall, R.: An explicit primitive conservative solver for the euler equations with arbitrary equation of state. Comput. Fluids 106340 (2024)
7. Barth, T., Jespersen, D.: The design and application of upwind schemes on unstructured meshes. In: 27th Aerospace Sciences Meeting. American Institute of Aeronautics and Astronautics (1989)

8. Bell, I.H., Wronski, J., Quoilin, S., Lemort, V.: Pure and pseudo-pure fluid thermophysical property evaluation and the open-source thermophysical property library coolprop. Ind. Eng. Chem. Res. **53**(6), 2498–2508 (2014)
9. Toro, E.F., Spruce, M., Speares, W.: Restoration of the contact surface in the HLL-riemann solver. Shock Waves **4**(1), 25–34 (1994)
10. Vinokur, M., Montagné, J.: Generalized flux-vector splitting and roe average for an equilibrium real gas. J. Comput. Phys. **89**(2), 276–300 (1990)
11. Baer, M.R., Nunziato, J.W.: A two-phase mixture theory for the deflagration-to-detonation transition (ddt) in reactive granular materials. Int. J. Multiphase Flow **12**(6), 861–889 (1986)
12. Re, B., Abgrall, R.: A pressure-based method for weakly compressible two-phase flows under a baer-nunziato type model with generic equations of state and pressure and velocity disequilibrium. Int. J. Numer. Methods Fluids **94**(8), 1183–1232 (2022)
13. Lochon, H., Daude, F., Galon, P., Hérard, J.M.: HLLC-type Riemann solver with approximated two-phase contact for the computation of the Baer-Nunziato two-fluid model. J. Comput, Phys (2016)
14. Chiocchetti, S., Müller, C.: A solver for stiff finite-rate relaxation in baer–nunziato two-phase flow models. In Lamanna, G., Tonini, S., Cossali, G.E., Weigand, B. (eds.), Droplet Interactions and Spray Processes, pp. 31–44. Springer International Publishing, Cham (2020)

[illegible]

Sonic Properties in Non-Ideal Flows

Marta Zocca and Alberto Guardone

Abstract The sonic properties of mono-component single-phase non-ideal flows are studied assuming steady, one-dimensional, inviscid, shock-free adiabatic flow. Under these hypotheses, the total enthalpy h^{t} and entropy s are uniform in the flow, and the sonic properties density ρ^*, speed of sound c^*, momentum density $\rho^* c^*$, and momentum flux density $\rho^* c^{*2}$ are readily computed from the total thermodynamic state specified by the total pressure and temperature $(P^{\mathrm{t}}, T^{\mathrm{t}})$. Application-relevant fluids and operating conditions are considered and isentropic expansions from reservoir to sonic conditions are investigated using both state-of-the-art thermodynamic models and the ideal-gas model. The evolution of selected thermodynamic properties along isentropic expansions is monitored and maps reporting the sonic properties in the P^{t}-T^{t} thermodynamic plane are constructed to illustrate the mechanism of dependence of the sonic properties on the total thermodynamic states. Several different non-ideal thermodynamic effects are identified and commented, which determine remarkable qualitative and quantitative differences between ideal and non-ideal flows.

Keywords NICFD · Mass flow rate · Chocked flows · Sonic properties · Nozzles · Injectors

M. Zocca (✉)
Fluid Dynamics, LES, LUT University, Lappeenranta, Finland
e-mail: marta.zocca@lut.fi

A. Guardone
Department of Aerospace Science and Technology, Politecnico di Milano, Milan, Italy
e-mail: alberto.guardone@polimi.it

A. Giauque et al. (eds.), *Proceedings of the 5th International Seminar on Non-Ideal Compressible Fluid Dynamics for Propulsion and Power*, ERCOFTAC Series 32,
https://doi.org/10.1007/978-3-031-92695-2_3

1 Introduction

One of the possible metrics used to quantify flow non-ideality is the so-called fundamental derivative of gasdynamics Γ [1], defined as:

$$\Gamma = \frac{1}{c}\left(\frac{\partial(\rho c)}{\partial \rho}\right)_s = 1 + \frac{\rho}{c}\left(\frac{\partial c}{\partial \rho}\right)_s \quad (1)$$

where c is the speed of sound, ρ is the density, and s is the specific entropy per unit mass. For perfect gases, and in general for flow regimes characterized by $\Gamma > 1$, the speed of sound c increases with density ρ. For a perfect gas, Eq. (1) reduces to $\Gamma = (\gamma + 1)/2$, which is always positive and greater than 1. For $\Gamma < 1$, the flow regime is termed non-ideal, and the speed of sound decreases with density.

In this work, the evolution of thermodynamic states from reservoir to sonic condition is studied over a wide range of operating conditions to investigate the influence reservoir conditions on application-relevant parameters such as the sonic momentum density $\rho^* c^*$ and the sonic momentum flux density $\rho^* c^{*2}$. The former property, $\rho^* c^*$, represents the maximum mass flow rate per unit area that a fluid-dynamic component can discharge starting from a given reservoir thermodynamic state $(P^{\mathrm{t}}, T^{\mathrm{t}})$. When the component is operated away from the design point, the sonic conditions attained at its minimum-area section change accordingly, and deviations from the design mass flow rate are observed. This is particularly relevant, for example, for Organic Rankine Cycles (ORC) [2] and supercritical CO_2 power cycles [3], where the chocked-flow conditions usually occurring in the turbine stator determine a specific operating point for the cycle. The latter property, $\rho^* c^{*2}$ is particularly relevant for state-of-the-art injectors, where fuel and oxidizer are mixed at pressures greater than the critical pressures of pure components [4]. The property $\rho^* c^{*2}$ influences the jet penetration depth in under-expanded conditions [5].

Differently from the ideal-gas case, the sonic properties of non-ideal flows are not in a relation of constant proportionality with total conditions. In other terms, the ratio of sonic to total properties is both fluid-dependent and condition-dependent. This concept is documented extensively, among others, in the work of Banholzer & Pfitzner [6], which compares the sonic momentum density and momentum flux density of Nitrogen and Methane as predicted by the ideal-gas and by the Redlich-Kwong Equation of State (EoS).

In this work, stagnation conditions located a) in the non-ideal thermodynamic region ($0 < \Gamma < 1$) of molecularly complex organic fluids and b) in the $\Gamma > 1$ region of fluids characterized by a lower molecular complexity are consisered. The corresponding sonic properties are computed by numerically solving the conservation equations, with fluid's thermodynamics described by the state-of-the-art EoS implemented in the CoolProp library [7]. The goal of this work is to identify the role of non-ideal thermodynamics in the variation of sonic properties with stagnation conditions. To this purpose, maps reporting the sonic properties on wide, application-relevant regions of the thermodynamic plane are constructed and the qualitative and quantitative behaviour of relevant flow properties is monitored along representative isentropic expansions.

The paper is structured as follows. In Sect. 2, the flow model is detailed and the non-ideal phenomena of interest in this work are introduced. Results are presented and discussed in Sect. 3 and in Sect. 4. Conclusions are reported in Sect. 5.

2 Sonic Properties of Ideal and Non-Ideal Isentropic Nozzle Flows

The sonic flow properties for chocked flows are computed by examining the quasi-one-dimensional steady expansion of a single-phase mono-component fluid from a reservoir with known thermodynamic state to an ambient at constant pressure. Such model is representative of the expansion occurring in a nozzle, whose geometry is of the de Laval type if the ambient pressure is such that a supersonic exit flow is entailed. Under the further assumptions of inviscid, adiabatic and shock-free flow, the governing equations reduce to the following well-known algebraic set of equations:

$$\rho u A(x) = \text{const.} \tag{2a}$$

$$h + \frac{1}{2}u^2 = \text{const.} \tag{2b}$$

$$s = \text{const.} \tag{2c}$$

where ρ is the density, u is the velocity module, $A(x)$ is the known cross-sectional area distribution of the nozzle, h is the static enthalpy per unit mass, s is the static entropy per unit mass. An Equation of State (EoS) completes the problem.

If the ideal-gas EoS is adopted, the enthalpy and the speed of sound depend only on temperature. Thus, the energy conservation equation leads directly to the well-known isentropic relations [8], which at the sonic condition ($M = 1$) evaluate to:

$$T^* = T^{\mathrm{t}}\left(\frac{2}{\gamma+1}\right), \qquad P^* = P^{\mathrm{t}}\left(\frac{2}{\gamma+1}\right)^{\frac{\gamma}{\gamma-1}}, \qquad \rho^* = \rho^{\mathrm{t}}\left(\frac{2}{\gamma+1}\right)^{\frac{1}{\gamma-1}} \tag{3}$$

Equation (3) expresses the well-known result that, for ideal-gas flows, the sonic and total thermodynamic properties are in a relation of constant proportionality. Since the value of the specific heat ratio γ is computed in the dilute-gas limit regardless of process conditions, the constants of proportionality in Eq. (3) depend only on the characteristics of the fluid.

If non-ideal thermodynamics is taken into account through a more complex EoS, enthalpy and speed of sound are in general function of two independent thermodynamic properties, and not of temperature only. Therefore, the proportionality relation between the total and sonic thermodynamic states no longer holds, and the solution to the nonlinear equation Eq. (2b) determines the thermodynamic state in sonic conditions. Assuming uniform total enthalpy h^{t} and entropy s, the flow properties along an expansion are defined by the solution of the following equation:

$$h^{\mathrm{t}} = h(\rho; h^{\mathrm{t}}, s) + \frac{1}{2}u^2(\rho; h^{\mathrm{t}}, s) \tag{4}$$

The only unknown in Eq. (4) is the density ρ, while the total enthalpy h^{t} and entropy s only act as parameters identifying stagnation conditions. For example, h^{t} and s can be directly computed from a known pair of total conditions $(P^{\mathrm{t}}, T^{\mathrm{t}})$ as $h^{\mathrm{t}} = h(P^{\mathrm{t}}, T^{\mathrm{t}})$ and $s = s(P^{\mathrm{t}}, T^{\mathrm{t}})$ using an appropriate EoS. The thermodynamic state in sonic conditions is obtained by solving the energy conservation equation Eq. (4) for the unknown ρ^*, by setting:

$$u(\rho^*; s, h^{\mathrm{t}}) = c(\rho^*; s, h^{\mathrm{t}}) \tag{5}$$

The chosen non-ideal EoS can then be used to compute other relevant thermodynamic properties from the pair (s, ρ^*). Once the sonic state is determined, the maximum mass flow rate is readily obtained as:

$$\dot{m} = \rho^* c^* \min_x A(x), \tag{6}$$

where the product $\rho^* c^*$ is the sonic momentum density.

The variation of the speed of sound along an isentropic expansion is expressed by the local value of the fundamental derivative of gasdynamics Γ, Eq. (1). The sonic speed of sound, which enters the definition of sonic momentum density and of the sonic momentum flux density, is thus strongly related to the evolution of Γ along the expansion. Integration of the two equivalent definitions of Γ included in Eq. (1) from sonic to total conditions highlights an explicit dependence of the sonic momentum density $\rho^* c^*$ and of the sonic speed of sound c^* on the local value of Γ along the expansion:

$$c^* = c^{\mathrm{t}} - \int_{\rho^*}^{\rho^{\mathrm{t}}} \frac{c(s, r)}{r} \left[\Gamma(s, r) - 1\right] \mathrm{d}r + \varphi(s) \tag{7a}$$

$$\rho^* c^* = \rho^{\mathrm{t}} c^{\mathrm{t}} - \int_{\rho^*}^{\rho^{\mathrm{t}}} c(s, r)\Gamma(s, r) dr + \phi(s) \tag{7b}$$

The above equations Eqs. (7a) and (7b) highlight the different mechanisms through which non-ideal thermodynamics influences the values of c^* and of $\rho^* c^*$. First, the extremes of the definite integrals (ρ^{t} and ρ^*) can both quantitatively depart from their ideal-gas counterparts. Second, the sign and value of Γ along the expansion determine the sign and value of the integrand functions in both Eqs. (7a) and (7b). In Eq. (7a), the integrand function is negative for all states characterized by $\Gamma < 1$, positive otherwise. Therefore, both qualitative and quantitative differences on the sonic speed of sound c^* of ideal and non-ideal expansions can be expected. If non-classical effects ($\Gamma < 0$) are not considered, Eq. (7b) suggests instead that only quantitative deviations between the ideal and non-ideal value of the sonic momentum density $\rho^* c^*$ are expected.

Notably, for ideal flows, Eqs. (7a) and (7b) reduce to:

$$\frac{c^*}{c^{\mathrm{t}}} = \left(\frac{\rho^*}{\rho^{\mathrm{t}}}\right)^{\frac{\gamma-1}{2}}, \qquad \frac{\rho^* c^*}{\rho^{\mathrm{t}} c^{\mathrm{t}}} = \left(\frac{\rho^*}{\rho^{\mathrm{t}}}\right)^{\frac{\gamma+1}{2}}, \tag{8}$$

which, consistently with Eq. (3), yield:

$$\frac{c^*}{c^{\mathrm{t}}} = \sqrt{\frac{2}{\gamma+1}}, \qquad \frac{\rho^* c^*}{\rho^{\mathrm{t}} c^{\mathrm{t}}} = \left(\frac{2}{\gamma+1}\right)^{\frac{(\gamma+1)}{2(\gamma-1)}} \tag{9}$$

3 Non-Ideal Thermodynamic Effects on the Sonic Properties

The evolution of relevant flow properties along selected isentropic expansions is now computed using both state-of-the-art EoS available from the CoolProp library [7] and the ideal-gas EoS to identify non-ideal thermodynamic effects. Two fluids exhibiting non-ideal thermodynamic effects are considered, namely the siloxane fluid MDM ($C_8H_{24}O_2Si_3$, $P_c = 14.100$ bar, $T_c = 564.09$ K) and Methane (CH_4, $P_c = 45.992$ bar, $T_c = 190.56$ K). The pressure and temperature ranges considered in this work are (0.1-$1.2P_c$, 0.8-$1.05T_c$) for MDM and (0.2-$17P_c$, 1.4-$2.6T_c$) for Methane. The distribution of the fundamental derivative of gasdynamics Γ, as computed by non-ideal EoS [9, 10] is represented as colored isolines on the P-T thermodynamic plane of Fig. 1 for MDM (left) and for Methane (right). The operating conditions of MDM considered here include portions of the superheated and of the supercritical regions, where both $\Gamma < 1$ and $\Gamma > 1$ regions are present and both non-ideal thermodynamics and non-ideal gasdynamics effects can be observed (see the classification proposed by Guardone et al. [11]). For Methane, the chosen operating range is fully supercritical in temperature and both subcritical and supercritical in pressure. The operating range conidered here is similar to the one investigated by Banholzer & Pfitzner [6]. Despite including only states characterized by $\Gamma > 1$, the results documented in Ref. [6] demonstrate that significant quantitative deviations between the predictions of ideal and non-ideal EoS are present, especially for low-temperature and high-pressure conditions.

In Fig. 1, selected isentropes are highlighted on the P-T thermodynamic plane for both MDM (left) and Methane (right). The reservoir conditions are identified as points A, B, and C on the diagrams, and their respective entropy levels are labelled as s_{A}, s_{B}, and s_{C}. For MDM, the three chosen expansions are representative of (A) a fully non-ideal expansion ($\Gamma < 1$) to sonic conditions in an ORC turbine stator [12], (B) a fully non-ideal expansion ($\Gamma < 1$) originating in the region where the lowest values of Γ are attained and (C) a transcritical expansion originating from the

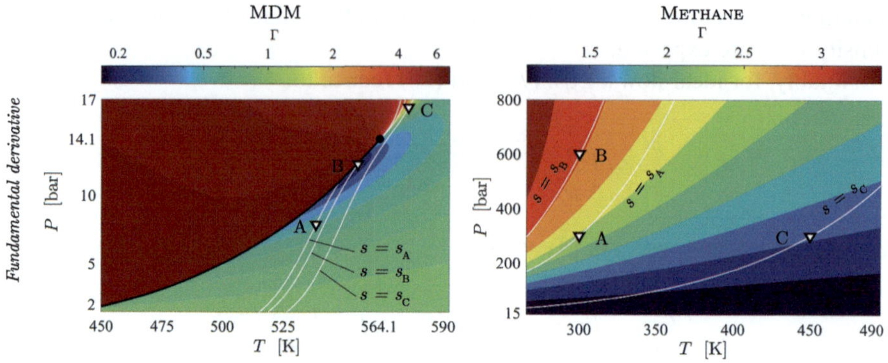

Fig. 1 Isolines of the fundamental derivative of gasdynamics in the P-T diagram for MDM (left) and for Methane (right). On each diagram, selected isentropes are highlighted, labelled as A, B, C

$\Gamma > 1$ region and terminating in the superheated vapour region, where $\Gamma < 1$. For Methane, the three expansions are representative of (A) low-pressure and (B) high-pressure injection processes of Compressed Natural Gas (CNG), see Ref. [4], and (C) a low-pressure, high-temperature expansion for which the ideal-gas behaviour is expected to be approached.

For all selected isentropes, the mass flux function $j(\rho; s, h^{\mathrm{t}}) = \rho u(\rho; s, h^{\mathrm{t}}) = \rho\sqrt{2(h^{\mathrm{t}} - h)}$ [13], the speed of sound $c(\rho; s, h^{\mathrm{t}})$, and the fundamental derivative of gasdynamics $\Gamma(\rho; s, h^{\mathrm{t}})$ along an expansion from the reservoir state $(P^{\mathrm{t}}, T^{\mathrm{t}})$ to an exhaust Mach number $M = 1.5$ are reported versus density ρ in Fig. 2. Notably, the mass flux function varies with density according to $(1/c)(\mathrm{d}j/\mathrm{d}\rho) = (M^2 - 1)/M$, which means that $j(\rho)$ increases (decreases) along an isentropic supersonic (subsonic) compression. The sonic conditions, which are marked as black dots along the $j(\rho)$, $c(\rho)$, and $\Gamma(\rho)$ curves of Fig. 2, correspond to the maxima of function $j(\rho)$, also see Ref. [14]. The maxima of $j(\rho)$ represent the momentum density in sonic conditions $\rho^* c^*$, while the sonic conditions marked on the speed of sound and fundamental derivative plots identify the sonic speed of sound c^* and sonic fundamental derivative Γ^*. In Fig. 2, continuous lines represent the evolution of j, c, and Γ computed by the non-ideal EoS Ref. [9, 10], while dotted lines represent the results obtained by the ideal-gas EoS. The total state $(P^{\mathrm{t}}, T^{\mathrm{t}})$ is the same for each 'Ideal' and 'Non-ideal' curve pair, therefore the observed differences are due to non-ideal thermodynamic effects.

The density levels attained along the considered isentropic expansions require careful consideration. From Fig. 2, large deviations are observed on the dimensional values of density, from total to exhaust, for both MDM and Methane. While for MDM the deviations are significant for all observed expansions, for methane deviations are prominent for expansions 'A' and 'B'. Indeed, the local values of Γ along expansions 'A' and 'B' significantly depart from the ideal-gas value $\Gamma = 1.1654$. For MDM, the total density ρ^{t} computed from the same input pair $(P^{\mathrm{t}}, T^{\mathrm{t}})$ is significantly underpredicted by the ideal-gas EoS for all the studied conditions. For Methane, ρ^{t} is

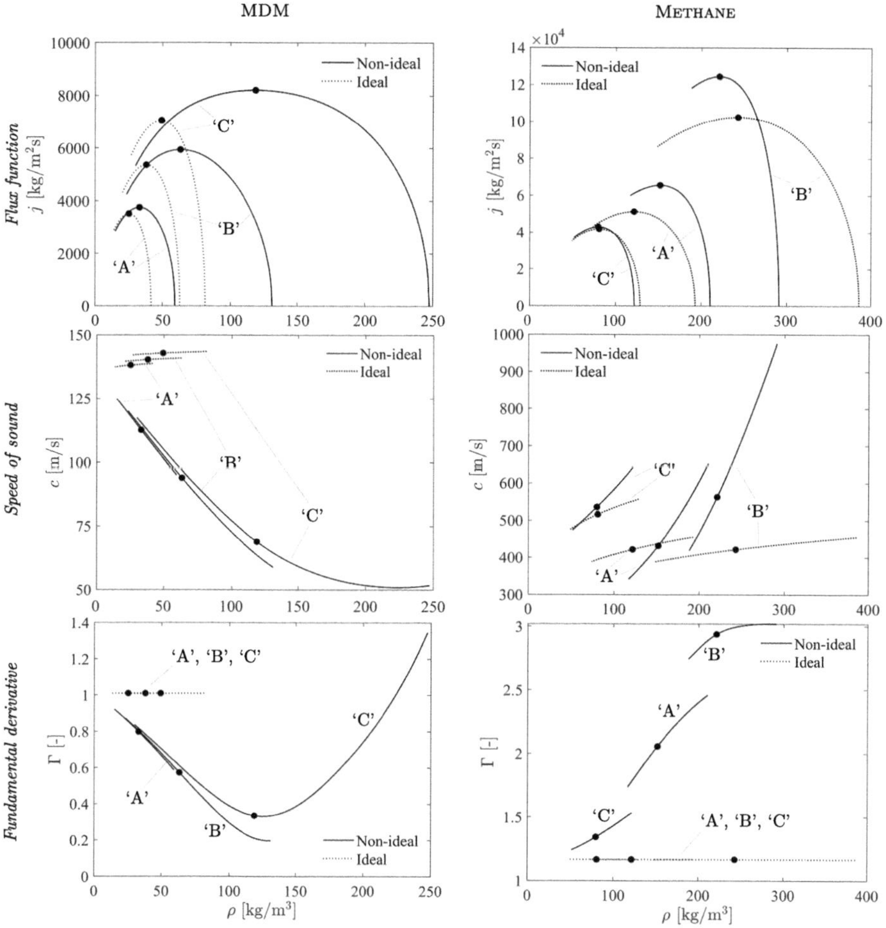

Fig. 2 Flux function (top), Speed of sound (middle), and Fundamental derivative of gasdynamics (bottom) as function of density. The fluid expands from reservoir to $M = 1.5$ along the isentropes A, B, C highlighted in Fig. 1. Results are reported for MDM (left) and for Methane (right). Non-ideal results (solid lines) are compared to ideal-gas results (dotted lines). Black dots represent sonic conditions

largely underpredicted by the ideal-gas EoS for expansion 'A', largely overpredicted for expansion 'B', and, as expected, only slightly overpredicted for expansion 'C'.

For MDM, the speed of sound in the total state c^t is remarkably overpredicted by the ideal-gas EoS for MDM (all conditions), underpredicted for Methane (all conditions). For both MDM and Methane, the ideal and non-ideal evolution of the speed of sound along the studied expansions are notably different. For Methane, $\Gamma > 1$ along all studied expansions and the speed of sound decreases along the expansions. The observed differences between ideal and non-ideal behavior are only quantitative. For MDM, instead, differences between ideal and non-ideal results are both quantitative and qualitative. Since $\Gamma < 1$ in all studied states besides at the beginning of expansion 'C', non-ideal results show a speed of sound increasing along all expansions (except in the proximity of ρ^{t} for expansion 'C'), while ideal results show a speed of sound increasing along all expansions. This is consistent with Eq. (7a).

For MDM, the values of the sonic momentum density (namely, the maxima of the flux function curves) predicted by ideal-gas and non-ideal thermodynamics are, at first glance, surprisingly close, despite the remarkable differences in the evolutions of density ρ and speed of sound c commented so far. The accuracy of the ideal-gas prediction is, however, the mere consequence of multiplying an underpredicted sonic density ρ^* by an overpredicted sonic speed of sound c^*. The same pattern was observed for several other organic fluids adopted in ORC applications, although results are not reported in this work for brevity. For Methane, instead, the ideal-gas model significantly underpredicts the sonic momentum density $\rho^* c^*$ for expansions 'A' and 'B', and, differently from MDM, no cancellation of opposite effects occurs.

Figure 3 reports the dimensional mass flux function j versus the static to total density ratio ρ/ρ^{t} for both MDM and Methane. From Fig. 3, it is apparent that, under the ideal-gas assumption, the static to total density ratio is independent of total conditions, while remarkable variations of ρ/ρ^{t} with the total state are observed if non-ideal thermodynamics is considered.

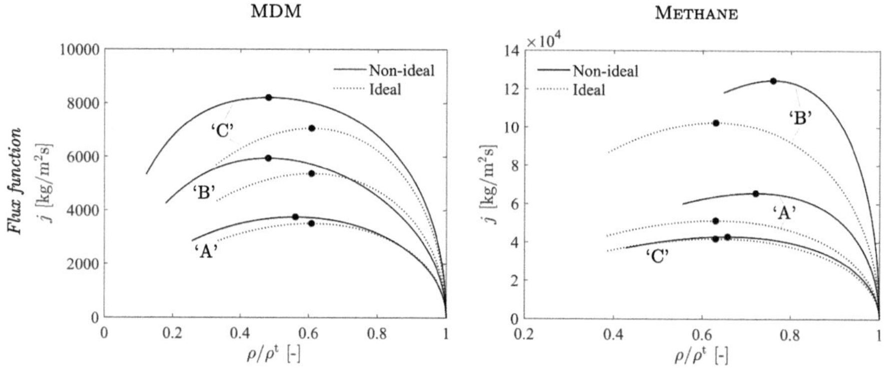

Fig. 3 Flux function as function of static to total density ratio. The fluid expands from reservoir to $M = 1.5$ along the isentropes A, B, C highlighted in Fig. 1. Results are reported for MDM (left) and for Methane (right). Non-ideal results (solid lines) are compared to ideal-gas results (dotted lines). Black dots represent sonic conditions

4 Sonic Momentum Density and Sonic Momentum Flux Density

In this section, the dependence of selected sonic properties on the total thermodynamic state is studied for a wide range of total conditions relevant to applications. The dependence of the sonic momentum density on total conditions is studied first for fluids MDM and Methane, under the same operating ranges considered in Sect. 3. The goal of this first analysis is to extend the observations made for selected isentropic expansions in Sect. 3 to the entire thermodynamic plane. Moreover, this analysis aims at quantifying the deviations between the ideal and non-ideal predictions of the sonic momentum density ρ^*c^* and of the sonic momentum flux density ρ^*c^{*2}. The analysis is then repeated for several other fluids: the sonic properties ρ^*c^* and ρ^*c^{*2} are carefully compared to identify possible common qualitative patterns in their dependence on stagnation conditions.

The maps reported in Figs. 4 and 5 show the distribution of the sonic properties density ρ^*, speed of sound c^* and momentum density ρ^*c^* in the P-T thermodynamic plane. The P-T values reported on the axes are the total pressure P^{t} and temperature T^{t}. The colored contour levels reported in Figs. 4 and 5 represent the sonic properties computed from the total thermodynamic state defined by the axes input pair (P^{t}, T^{t}). Sonic properties are computed using either the non-ideal EoS [9, 10] (ρ^*, c^*, ρ^*c^*) or the ideal-gas EoS ($\rho^*_{\mathrm{id.}}$, $c^*_{\mathrm{id.}}$, $(\rho^*c^*)_{\mathrm{id.}}$). Expansions determining the occurrence of a second phase in the flow are excluded from the analysis.

For both MDM and Methane, the distribution of the sonic properties over the P-T plane confirm the trends discussed for Figs. 2 and 3. Although quantitative differences are observed for all computed sonic properties, only the sonic speed of

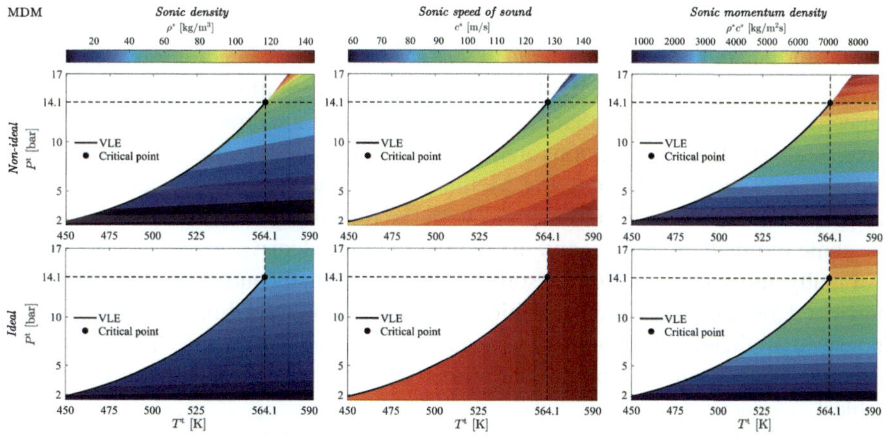

Fig. 4 Maps of sonic density, sonic speed of sound, and sonic momentum density in the P-T plane for MDM. The (P, T) pairs along the axes represent the total conditions. Results are reported for non-ideal model (top row) and for ideal-gas model (bottom row)

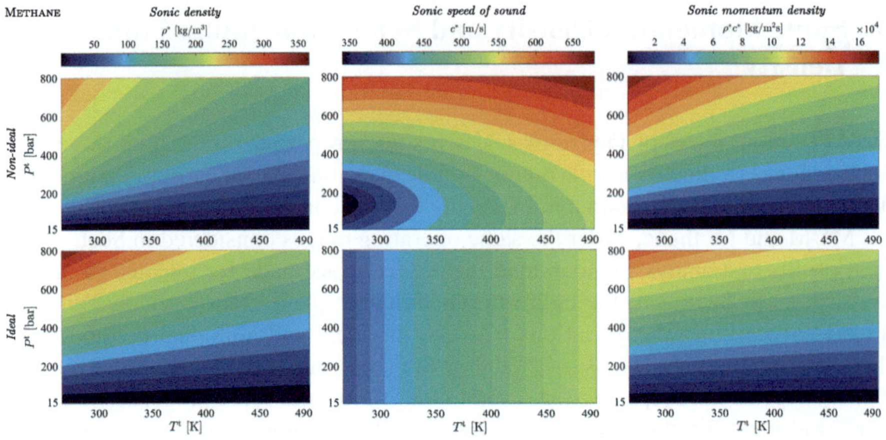

Fig. 5 Maps of sonic density, sonic speed of sound, and sonic momentum density in the P-T diagram for Methane. The (P, T) pairs along the axes represent the total conditions. Results are reported for non-ideal model (top row) and for ideal-gas model (bottom row)

sound displays qualitative differences between ideal and non-ideal results. The sonic speed of sound c^* computed using the non-ideal model shows a strong dependence on both T^{t} and P^{t}. The lowest values are observed in the region close to the critical point. As opposite, $c^*_{\mathrm{id.}}$ is function of temperature only.

To quantify the deviation between ideal-gas and non-ideal predictions for both MDM and Methane, the relative sonic momentum density $(\rho^*c^*)/(\rho^*c^*)_{\mathrm{id.}}$ and the relative sonic momentum flux density $(\rho^*c^{*2})/(\rho^*c^{*2})_{\mathrm{id.}}$ are displayed over the P-T plane in Fig. 6. In the studied P-T range, (ρ^*c^*) is generally underpredicted by the ideal-gas model for both MDM and Methane. The largest deviations are observed in the supercritical region for MDM and in the low-temperature, mid-pressure region of the studied P-T range for Methane. (ρ^*c^{*2}), instead, is generally overpredicted by the ideal-gas model for MDM. Again, the largest deviations are observed in the supercritical region. For methane, (ρ^*c^{*2}) is generally underpredicted by the ideal-gas model. The largest deviations are observed at the low-temperature, high-pressure end of the studied P-T range. For Methane, the deviation between ideal and non-ideal results displayed in Fig. 6 agrees with the results reported in Ref. [6] for both the sonic momentum density and the momentum flux density.

Finally, the dependence of the sonic momentum density and of the sonic momentum flux density on stagnation conditions is studied for several fluids relevant to either industrial applications or to ongoing academic research within the NICFD community. The name, formula, and critical properties of the studied fluids are reported in Table 1.

For all fluids considered here, the operating range is defined in terms of reduced variables $T_r = T/T_c$ and $P_r = P/P_c$, where T_c and P_c are the fluid's critical temperature and pressure, respectively. Molecularly complex fluids MDM, R134A, D6, and NOVEC649 (Fluid set 1) are studied in the range $T_r = 0.8$-1.05, $P_r = 0.1$-1.2,

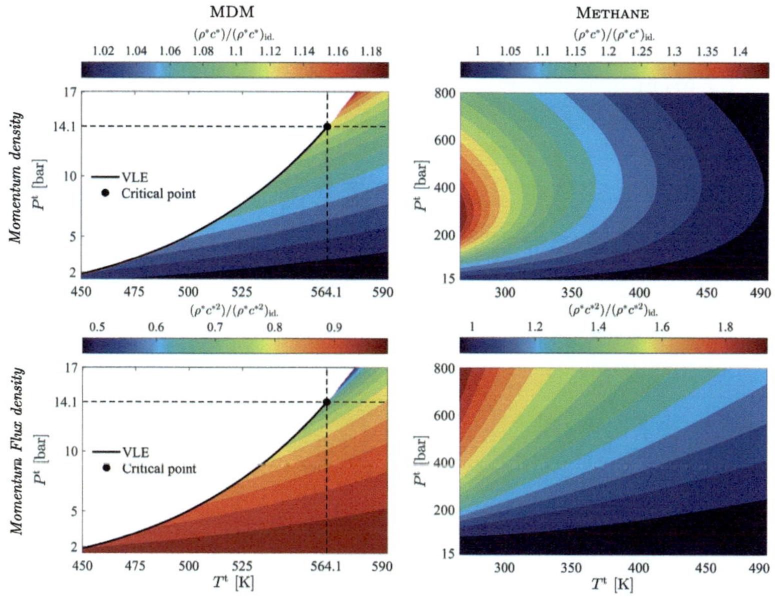

Fig. 6 Ratios of non-ideal to ideal-gas sonic momentum density (top row) and sonic momentum flux density (bottom row) in the (P, T) plane for MDM (left) and Methane (right). The (P, T) pairs along the axes represent the total conditions

Table 1 Critical properties of the studied fluids

Fluid name	Formula	T_c [K]	P_c [Pa]	ρ_c [kg/m^3]
MDM	$C_8H_{24}O_2Si_3$	564.09	1410045	256.74
R134A	$C_2F_4H_2$	374.21	4059280	511.90
D6	$C_{12}H_{36}O_6Si_6$	645.78	961000	279.10
NOVEC649	$C_6F_{12}O$	441.81	1869027	606.80
Nitrogen	N_2	126.19	3395800	313.30
Hydrogen	H_2	33.15	1296400	31.26
Methane	CH_4	190.56	4599200	162.66
Carbon dioxide	CO_2	304.13	7377300	467.60

where they all exhibit large regions characterized by $0 < \Gamma < 1$. The less complex compounds Nitrogen, Hydrogen, Methane, Carbon dioxide (Fluid set 2) are studied in the range $T_r = 1.5$-3, $P_r = 0.2$-20, where Γ is everythere > 1. For all fluids, expansions determining the occurrence of a second phase in the flow are excluded from the analysis.

In Fig. 7, the sonic momentum density and the sonic momentum flux density, made nondimensional with the fluid's critical properties as $\rho^* c^* / \sqrt{\rho_c P_c}$ and $\rho^* c^{*2} / P_c$, are mapped versus the reduced total temperature $T_r^{\rm t}$ and the reduced total pressure $P_r^{\rm t}$.

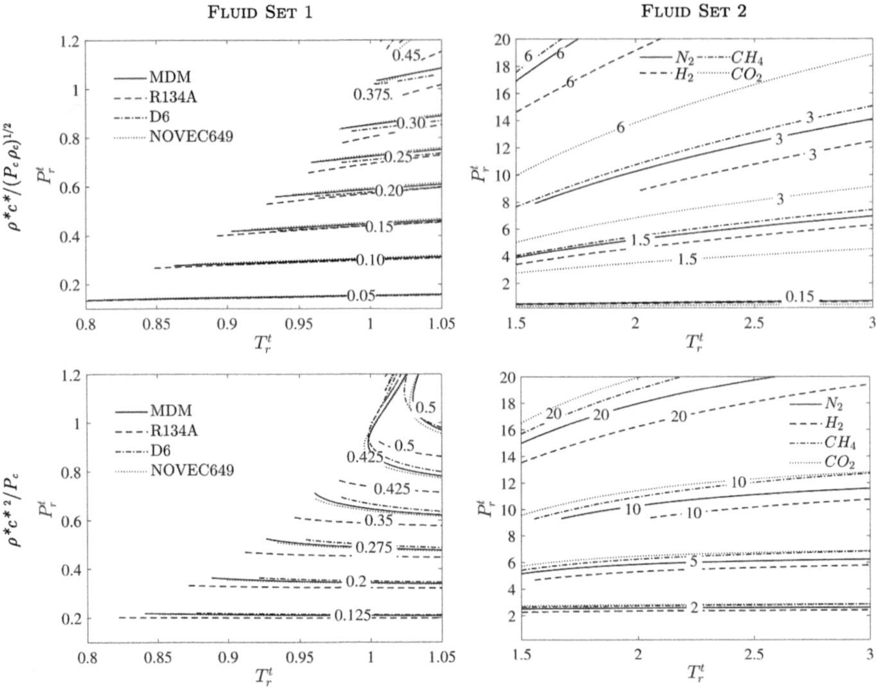

Fig. 7 Maps of sonic momentum density (top row) and sonic momentum flux density (bottom row), made nondimensional with the fluid's critical properties as $\rho^*c^*/\sqrt{\rho_c P_c}$ and ρ^*c^{*2}/P_c. Results are reported for Fluid set 1 (left) and for Fluid set 2 (right). Fluid properties are given in Table 1

The dependence of $\rho^*c^*/\sqrt{\rho_c P_c}$ and ρ^*c^{*2}/P_c on T_r and P_r is at least qualitatively similar among the two studied sets of fluids. For Fluid set 1, $\rho^*c^*/\sqrt{\rho_c P_c}$ is mostly governed by P_r^{t} for total pressures below the fluid's critical pressure. At higher total pressures, an increasingly stronger dependence of $\rho^*c^*/\sqrt{\rho_c P_c}$ on T_r^{t} is observed. Similar considerations apply to ρ^*c^{*2}/P_c. Remarkably, a partial quantitative correspondence between the values of $\rho^*c^*/\sqrt{\rho_c P_c}$ for different fluids is observed in the low-pressure end of the studied operating range. For Fluid set 2, both $\rho^*c^*/\sqrt{\rho_c P_c}$ and ρ^*c^{*2}/P_c are mostly governed by P_r^{t} in the low-pressure end of the studied range. At higher total pressures, an increasingly stronger dependence on T_r^{t} is observed. No quantitative correspondence is found between the values of $\rho^*c^*/\sqrt{\rho_c P_c}$ and of $\rho^*c^*/\sqrt{\rho_c P_c}$ of different fluids among Fluid set 2.

5 Conclusions

The sonic properties of non-ideal isentropic expansions are studied, with particular focus on the momentum density and on the momentum flux density, due to their relevance for industrial applications. Similarly to ideal-gas flows, sonic conditions of non-ideal flows are shown to be largely dependent on the total thermodynamic state. For both ideal and non-ideal flows, sonic properties embed a dependence on the total thermodynamic states through the total enthalpy and entropy entering the energy conservation equation. While for non-ideal flows the total thermodynamic state determines the qualitative and quantitative evolution of the flow properties along the expansion, for ideal flows the total and sonic thermodynamic states are in a relation of proportionality, regardless of the total thermodynamic state. Moreover, large deviations between the total thermodynamic properties derived from the total property pair $(P^{\mathrm{t}}, T^{\mathrm{t}})$ are present for all total states characterized by a value of the fundamental derivative of gasdynamics Γ departing from the ideal-gas value $\Gamma = (\gamma + 1)/2$. The dependence of the sonic properties on the total thermodynamic state was studied for several fluids and over a wide range of operating conditions. Common qualitative patterns in the behaviour of different fluids were identified by comparing the sonic properties in terms of reduced variables.

Acknowledgements This research was funded by the Research Council of Finland under grant number 342135, project *Non-ideal compressible flows in turbomachinery*.

References

1. Thompson, P.A.: A fundamental derivative in gasdynamics. Phys. Fluids **14**(9), 1843–1849 (1971)
2. Ennio, M., Marco, A.: Organic Rankine Cycle (ORC) power systems: Technologies and applications. 09 (2016)
3. Guo, J.-Q., Li, M.-J., He, Y.-L., Jiang, T., Ma, T., Jin-Liang, X., Cao, F.: A systematic review of supercritical carbon dioxide(S-CO2) power cycle for energy industries: Technologies, key issues, and potential prospects. Energy Conv. Manag. **258**, 115437 (2022)
4. Christoph, T., Matthias, B., Michael, P.: Real-gas effects and phase separation in underexpanded Jets at Engine-relevant conditions. In: 2018 AIAA Aerospace Sciences Meeting (2018)
5. Drozda, T.G., Robert, B.: CFD analysis of mixing characteristics of several fuel injectors at hypervelocity flow conditions. In: 52nd AIAA/SAE/ASEE Joint Propulsion Conference (2018)
6. Banholzer, M., Pfitzner, M.: Correlations for the choked mass and momentum flux density considering real-gas thermodynamics. ASME J. Fluids Eng. **141**(8), 081202 (2019)
7. Bell, I.H., Wronski, J., Quoilin, S., Lemort, V.: Pure and pseudo-pure fluid thermophysical property evaluation and the open-source thermophysical property library CoolProp. Ind. Eng. Chem. Res. **53**(6), 2498–2508 (2014)
8. Thompson, P.A.: Compressilbe-fluid dynamics. McGraw-Hill (1988)
9. Thol, M., Dubberke, F.H., Baumhögger, E., Vrabec, J., Span, R.: Speed of sound measurements and fundamental equations of state for octamethyltrisiloxane and decamethyltetrasiloxane. J. Chem. Eng. Data **62**(9), 2633–2648 (2017)
10. Setzmann, U., Wagner, W.: A new equation of state and tables of thermodynamic properties for methane covering the range from the melting line to 625 K at pressures up to 1000 MPa. J. Phys. Chem. Ref. Data **20**(6), 1061–1155 (1991)
11. Guardone, A., Colonna, P., Pini, M., Spinelli, A.: Nonideal compressible fluid dynamics of dense vapors and supercritical fluids. Ann. Rev. Fluid Mech. **56**, 241–269 (2024)
12. Turunen-Saaresti, T., Uusitalo, A., Honkatukia, J.: Design and testing of high temperature micro-ORC test stand using Siloxane as working fluid. J. Phys. Conf. Ser. **821**(1), 012024 (2017)

13. Kluwick, A.: Transonic nozzle flow of dense gases. J. Fluid Mech. **247**, 661–688 (1993)
14. Guardone, A., Vimercati, D.: Exact solutions to non-classical steady nozzle flows of BetheZeldovichThompson fluids. J. Fluid Mech. **800**, 278–306 (2016)

Start-Up Waves in Shock-Capturing Schemes for Ideal and Non-Ideal Compressible Flows

Marta Pederzolli and Alberto Guardone

Abstract Spurious start-up waves are studied for one-dimensional compressible flows in both ideal and non-ideal thermodynamic states. Numerical simulations are carried out on simple one-dimensional problems using both a Godunov scheme, the Roe scheme, and the Davis scheme. Numerical experiments show that two start-up waves are generated from an initial single shock wave for the ideal, dilute gas case. These are shown to correspond to a contact surface and an acoustic wave of the family opposed to the one pertaining to the shock. If the initial discontinuity is a contact discontinuity, no start-up waves are observed. In the non-ideal case, a similar outcome is observed for shock waves. Unlike the ideal case, contact discontinuities produce two start-up waves, one for each acoustic field, moving away from the contact discontinuity in both directions. This empirical observation is explained by the non-linearity of the jump relations across a contact discontinuity in the nonideal cases, whereby a linear relation exists linking conservative variables across a contact discontinuity in the ideal regime.

Keywords Compressible flows · Shock-capturing schemes · Start-up waves · Monotonicity · Non-Ideal compressible fluid dynamics

1 Introduction

The numerical solution of Conservation Laws (CLAWs) is an active field of research due to the diverse, relevant applications, most notably, the fluid dynamics of compressible flows [1]. Possibly, the major difficulty in solving CLAWs is the formation of non-linear discontinuities, usually referred to as shock waves borrowing from the

M. Pederzolli (✉) · A. Guardone
Politecnico di Milano, Milan, Italy
e-mail: marta.pederzolli@polimi.it

A. Guardone
e-mail: alberto.guardone@polimi.it

A. Giauque et al. (eds.), *Proceedings of the 5th International Seminar on Non-Ideal Compressible Fluid Dynamics for Propulsion and Power*, ERCOFTAC Series 32,
https://doi.org/10.1007/978-3-031-92695-2_4

nomenclature well-established in gasdynamics. Two major approaches have been devised to solve CLAW embedding discontinuities: *shock-fitting* methods and *shock -capturing* methods. In the first technique, the trajectory of the discontinuity is fitted, and the solution is computed before and after it. In the latter, instead, the goal is to capture discontinuities in the solution as a smooth flow feature. This is also the most popular method, due to the geometrical complexity of the shock front. Several methods and theoretical results are available in the literature, but unfortunately, those are only applicable to the scalar case. In particular, there is no guarantee that numerical schemes preserving the monotonicity of the solution for scalar CLAW apply to systems of conservation law, even in one dimension. A notable example is the simple case of an initial discontinuity propagating in one dimension. Even if the initial discontinuity corresponds to the exact solution of a single shock wave, oscillations—termed *start-up waves*—are generated by all shock capturing schemes as the shock settles into its numerical viscous profile [2, 3]. Different hypotheses have been made by other authors: Karni [4] investigated the first-order upwind scheme and the Lax–Friedrichs scheme, comparing the phase space plot of the solutions with plots of the modified equations. Xu and Hu [5] analyzed the dynamical effects in the projection stage of Godunov-type schemes for the nonlinear system, concluding this stage gives rise to numerical issues. Pantano [6] stated the cause is the violation of mechanical equilibrium at density interfaces of flows with real gas EOS, and presented a possible cure for the van der Waals model. Arora and Roe [7] and Zaide [8] provided a convincing explanation of the failure of shock-capturing schemes to solve the gasdynamic equations, which is based on physical grounds: "Slowly moving shockwaves will generate spurious oscillations in a code employing either the Godunov or Roe fluxes whenever the Hugoniot curves are not straight lines in the phase space of conserved variables. For other fluxes giving rise to one-point stationary shocks, the locus that must be a straight line is the locus traced by that intermediate point". This work studies start-up waves for one-dimensional compressible flows in both ideal and non-ideal thermodynamic states, following the latter interpretation [7]. Non-ideal thermodynamic conditions near the liquid-vapour saturation curve and critical point are modelled by the simple van der Waals equation of state under the assumption of constant specific heat at constant volume. Numerical simulations are carried out on simple one-dimensional problems using a Godunov scheme—the Roe scheme—and the Davis scheme.

2 Problem Statement

This study uses one-dimensional compressible Euler equations. These are a system of nonlinear hyperbolic conservation laws that govern the dynamics of a compressible material, for which the effects of volume forces, viscosity, and heat are neglected. Computing solutions containing discontinuities—such as shockwaves—pose stringent requirements on both the mathematical formulation and the numerical schemes used to solve the equations. Formulations based on non-conservative variables fail to compute the correct solution in the case of discontinuities, since they give the wrong jump conditions, resulting in an inaccurate shock speed. In this work, governing equations are written in terms of *conservative variables*—mass density $\rho(x, t)$, momentum density $\rho u(x, t)$, and the total energy density per unit volume E_{tot}—to guarantee the correct evaluation of the shock speed. Moreover, *conservative methods* are employed, which guarantee conservation at the discrete level: the *Roe upwind scheme* [9], an Approximate Riemann solver of Godunov type, and the *Davis scheme* [10]. In Fig. 1a are shown several cells within which there is a shock. The discrete representation of the wave on the grid is shown by the dots in each cell. Advancing in time, the discontinuity will move in different positions, and the numerical value inside the cell comprising it will become an intermediate point between the initial and final states on the discrete grids. The numerical schemes considered compute such intermediate points in a conservative manner: either by Godunov average or by conservative sampling. When solving for initial data representing a shock, the exact solution is a single shock with no anomalies generated. While solving the problem numerically, the wave moves and enters one of the cells, as shown in Fig. 1a. At the first time step, due to the Godunov average (or to conservative sampling, in the case of Davis method) the solution has an intermediate point inside the cell that is now containing the shock wave: such value, due to conservation, is a linear combination of the initial and final conditions and therefore lies on the straight line connecting these two in the conservative variables phase space. Since the shock adiabats are curved lines, such intermediate state does not lie on any of the two curves, as shown in Fig. 1b . Therefore, at the second time step, two Riemann problems will be solved at the two interfaces of the cell where the shock is. However, the resulting intermediate states at the following time step are not a linear combination of the initial and final states like before: they are a linear combination of intermediate states that solve a nonlinear problem. This implies that these do not lie on the straight line connecting the pre-shock and post-shock conditions in the conservative variables phase space.

3 Results

The investigation on start-up waves is carried out using a one-dimensional solver for Euler equations, with discontinuous initial data [11]. Initial data generating a single shock-wave moving fast across the grid is chosen. In the second part of this section, the same phenomenon is investigated for contact discontinuities, modeling the flow both as a Perfect Ideal Gas (P.I.G.) and then with the simple van der Waals model to account for non-ideal effects. For the whole discussion, the variables are to be considered non-dimensional. The scaling is made using values at the liquid-vapour critical point.

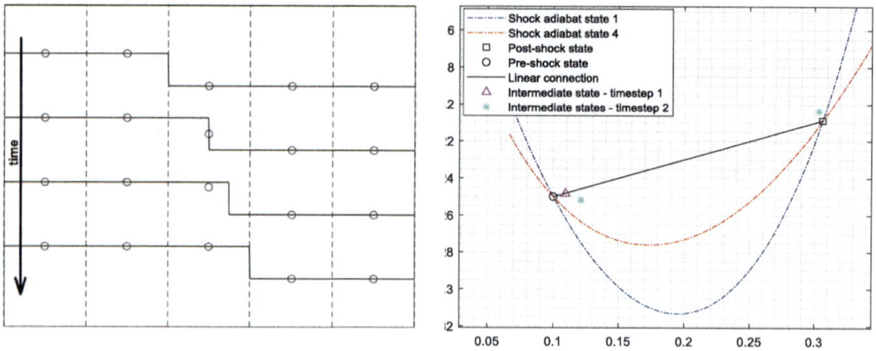

(a) (b)

Fig. 1 In Fig. 1a are shown several cells (dashed lines) within which there is a shock moving to the right (solid line). The numerical representation of the shock on the grid is given by cell-centered values, here reported with dots. This can be interpreted as a figure of the shock moving to the right, captured at four different times. On the right (Fig. 1b) the initial and final states are shown together with the intermediate state evaluated by the solver at the first time step, and the intermediate states at the second time step, in the conservative variables phase space. The latter do not lie on the black straight line, since they are linear combinations of other intermediate states that are a solution of a non-linear problem

Table 1 Starting set of data for the generation of a single, fast-moving shockwave

ρ_L	u_L	P_L	ρ_R	u_R	P_R
0.306088173547956	4.31790123	0.590833333333416	0.1	2.5	0.1

3.1 *Single Shock-Wave in Ideal Flow*

The data for this test case are taken from the paper by Arora and Roe [7] on post-shock oscillations, and it is the same in pressure and density, while the shock speed is increased by adding 10 to both left and right velocities. The precise values are written in Table 1: since the number of digits provided in [7] is not sufficient to distinguish numerical artifacts properly, in the present work an adjusted set is used to guarantee that the two points lies along the Hugoniot line one of the other up to machine precision. This is a set of initial conditions to simulate a single shock wave with 1D full Euler equations for a gas. The simulation is performed with 1 600 points in the spatial grid, with a $\Delta t = 10^{-4}$ and 800 iterations in time. The shock speed was increased with respect to the reference to avoid the generation of spurious oscillations and to observe the generation of start-up waves only. Results are shown in Fig. 2.

The scheme used for the numerical solution is a simple first-order upwind to show that the problem presents itself even when higher-order methods are not involved. The simulation is also performed at the same time with Davis scheme, which does

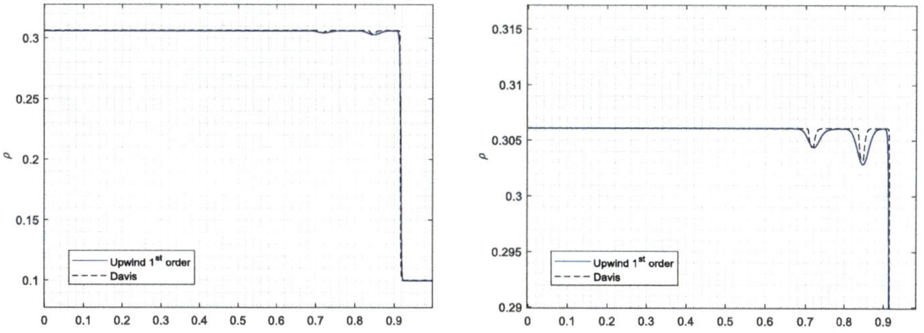

(a) *Density*

(b) *Density zoom on post-shock area*

Fig. 2 Plot of the results for the solution at time $t = 0.08$ s to the 1D Euler equations with primitive variable data ($\mathbf{u} = (\rho, u, p)^T$) from Table 1 using a first-order upwind scheme and Davis scheme. Only the density result is reported for clarity. The hypothesis of ideal gas is employed. The resulting discontinuity is a fast-moving single shock wave

not require a Riemann solver. This choice was made to show that start-up waves are not strictly related to the type of solver employed. Figure 2 shows the results for the density evaluated in the space grid. On the right, the zoom highlights the post-shock area of the graph where start-up waves can be observed. The start-up wave on the left corresponds to the genuinely non-linear type-1 characteristic field; the one between the shock and the first start-up wave is associated to the type-2 linearly degenerate field, as shown in the following analysis. The phenomenon is visible through the isolines of density and pressure in the $x - t$ plane, shown in Fig. 3.

Lines representing the initial shock wave (type-3 wave) appear very clearly. In the density isolines, two more waves appear in the picture, confirming the type-1

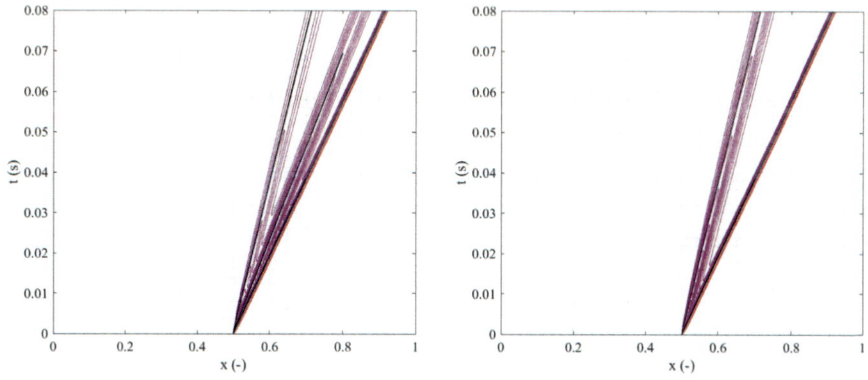

(a) *Density isolines*

(b) *Pressure isolines*

Fig. 3 Isolines of density and pressure in the x-t plane for the fast-moving single shock from Fig. 2. The black lines are the exact characteristic waves plotted for comparison

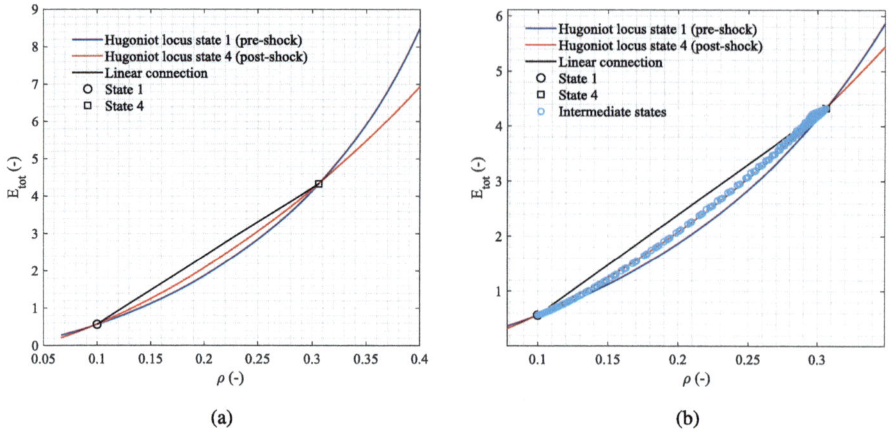

Fig. 4 Plot of pre-shock and post-shock states with the respective shock adiabats in conserved variables phase space for the ideal fast-moving shock wave case. In Fig. 4b, the blue dots represent the intermediate states

(compression acoustic wave) and type-2 (contact discontinuity) nature of the start-up waves. The type-2 wave is not present in the pressure isolines, as expected, since it is a contact discontinuity.

A final understanding comes from the plots of flow states in the conserved variables phase space, shown in Fig. 4.

As it is well known, in the case of a shockwave the shock adiabat is a curved line in the conserved variables phase space. This is consistent with Arora and Roe [7] result: when the locus is not a straight line, the generation of start-up waves is to be expected. In Fig. 4b the blue dots represent the intermediate states in the profile of the shock at different time steps, showing that the shock settles on a profile that significantly differs from the ones given by the two shock adiabats.

3.2 Contact Discontinuity in Ideal and Non Ideal Flow

The second type of discontinuity analyzed in this work is a contact discontinuity, where density is the only variable changing between the three considered. The test case used is reported in Table 2.

Table 2 Starting set of data for the generation of an ideal contact discontinuity

ρ_L	u_L	P_L	ρ_R	u_R	P_R
1.4	0.2	1.0	1.0	0.2	1.0

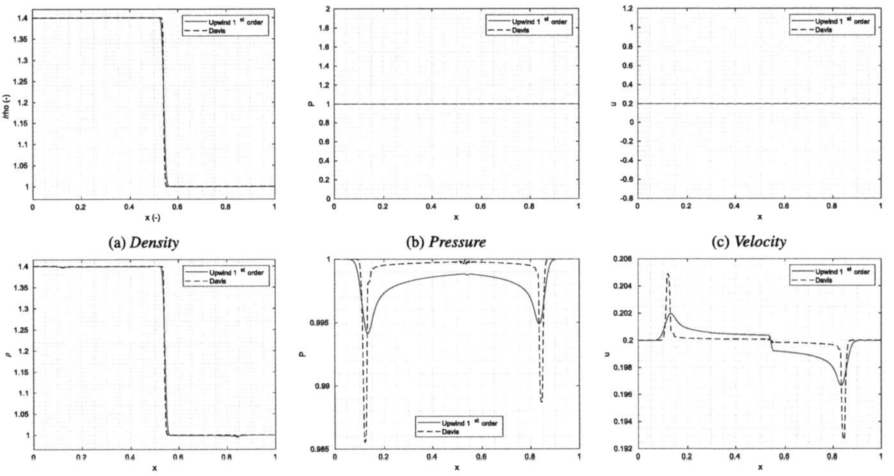

(a) *Density* (b) *Pressure* (c) *Velocity*

(d) *Density* (e) *Pressure* (f) *Velocity*

Fig. 5 Plot of the results for the solution to the 1D Euler equations with primitive variable data ($\mathbf{u} = (\rho, u, p)^T$) from Table 2 using a first-order upwind scheme and Davis scheme. The hypothesis of ideal gas is employed in the first case (Fig. 5a, b, c), while in the others the simple model of Van der Waals is employed. The resulting discontinuity is a contact discontinuity

This case is quite enlightening about the root cause of start-up waves. Both a test with P.I.G. model and a non ideal one are run, with fundamentally different results. Non-ideal thermodynamic conditions in the proximity of the liquid-vapor saturation curve and critical point are modeled by the simple van der Waals equation of state under the assumption of constant isochoric specific heat. The specific heat at constant pressure varies from state to state. The hypothesis of polytropic, constant isochoric specific heat, gas is made for simplicity. The initial data are the same in both the ideal and non-ideal contact discontinuity case, written in Table 2. In both cases, the simulation runs with a grid with 1 000 points, a $\Delta t = 10^{-4}$, and for 2 000 iterations. Results are shown in Fig. 5.

In the ideal gas case, as expected, the density is the only parameter changing, and it does so without any start-up wave. When the non-ideal gas model is employed, instead, unexpected numerical artifacts appear. In particular, in the density results shown in Fig. 5d, two start-up waves, one on the left side and one on the right side of the discontinuity, are generated. While this phenomenon is well known for shock waves, its appearance in contact discontinuities can bring some clarity about the underlying cause. These waves also affect the other two variables in a way not seen in the previous cases.

When plotting the isolines in the $x - t$ plane–shown in Fig. 6–the results are quite different: while in the ideal case the only visible wave is the actual contact discontinuity, in the non-ideal the two start-up waves (one left travelling and one right travelling) appear, confirming the observations made in Fig. 5. These are type-

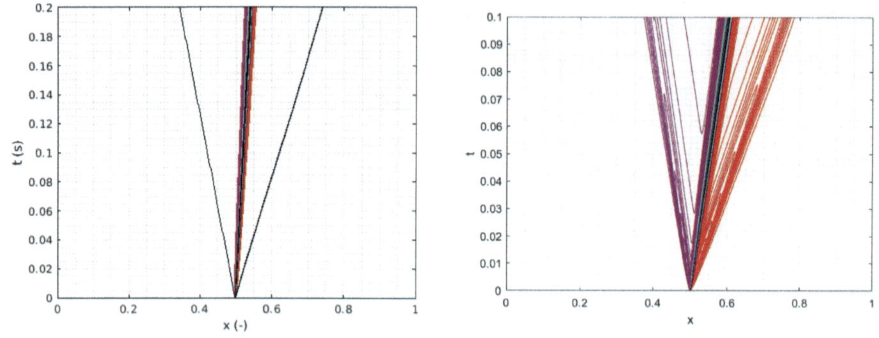

(a) *Density isolines - ideal gas* (b) *Density isolines - non ideal gas*

Fig. 6 Isolines of density in the x-t plane for the ideal (Fig. 6a) and non-ideal (Fig. 6b) contact discontinuity from Fig. 5

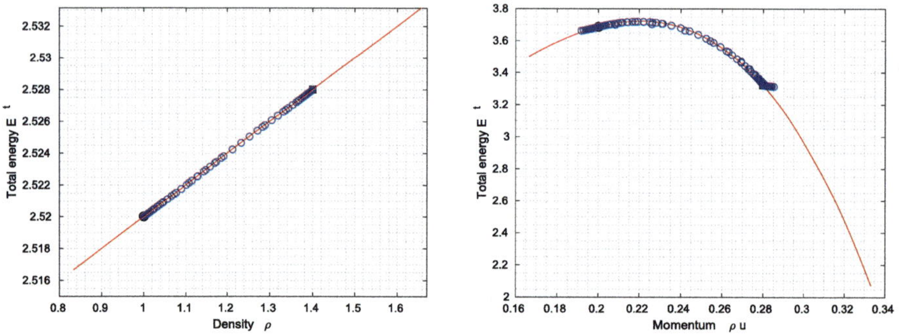

(a) (b)

Fig. 7 Plot of pre-discontinuity and post-discontinuity states with the respective Hugoniot locus in conserved variables phase space for the ideal (Fig. 7a) and non ideal (Fig. 7b) contact discontinuity case. Blue dots represent the intermediate states

1 and type-3 acoustic waves, which have their own characteristic speed. This result looks quite trivial, but it is fundamental to understand the actual reason behind the generation of numerical oscillations: in fact, it confirms the initial hypothesis, since for a contact discontinuity in the ideal gas case, the Hugoniot curve is a straight line. Therefore, the solution of each Riemann problem lies exactly on the same line and does not generate any spurious wave.

The results are finally shown in the conserved-variables phase-space plane in Fig. 7. The contact discontinuity (2-wave) is, in the ideal case, a linearly degenerate field. Therefore, the Hugoniot locus for the 2-wave is linear, not generating any start-up wave or spurious oscillation, since all intermediate states lie on that straight line. Figure 7a also shows that all the intermediate states lie on the straight line connecting the initial and final states, confirming again that the method without generation of start-up waves is always conservative, even locally. Instead, as shown in Fig. 7b, the Hugoniot locus in the case of a *non-ideal gas* is not linear: it is a curve. This finally confirms the initial hypothesis about the generation of start-up waves while extending it: they generate in the case of *any* kind of discontinuity if the Hugoniot curve is not straight when using shock-capturing methods. The last analysis to be carried out is the one about the intermediate states, plotted in the conserved variable phase space to gain a final understanding about the resolution path of a shock-capturing method in this case. Just like in the case of a shock wave, in Fig. 7b the intermediate points of the solution at different time steps do not lie on the straight line connecting the initial and final states.

4 Conclusions

The generation of start-up waves in numerical simulations of discontinuities, with a one-dimensional solver of Euler equations, was studied considering different test cases with both shock waves and contact discontinuities. This was first done using the ideal gas equation of state, like in [7] and [12], giving the expected results and visualizing in a better way the root cause of the problem. The original contribution of this work is given by the analysis of the same problem in a non-ideal regime: this is obtained by using the van der Waals equation of state, under the hypothesis of polytropic gas for simplicity.

The contact discontinuity example is the one that brings some light on the matter: even if this kind of discontinuity is considered linearly degenerate, in the non-ideal case its Hugoniot curve is *not* linear, giving rise to the same phenomena that was always considered a peculiarity of shock waves.

Start-up waves and spurious oscillations are *always* generated whenever a shock-capturing method is employed and the Hugoniot curves are not straight lines. This also includes methods that are different from Godunov or Roe fluxes, such as the Davis method—which gives the same kind of results in all the cases evaluated here. The discovery made in this work is that start-up waves are not limited to genuinely non-linear characteristic fields (shock waves of type 1 and 3 in gasdynamics) but also to linearly degenerate fields (type-2 contact discontinuities) if the Rankine-Hugoniot locus is not a straight line in the conservative-variable space. This is indeed the case for a non-ideal equation of state, such as the simple van der Waals model.

References

1. Randall, J.L.: Numerical methods for conservation laws. Lectures in Mathematics. ETH Zürich, Birkhäuser Basel (1992)
2. Woodward, P., Colella, P.: The numerical simulation of two-dimensional fluid flow with strong shocks. J. Comput. Phys. **54**(1), 115–173 (1984)

3. Roberts, T.W.: The behavior of flux difference splitting schemes near slowly moving shock waves. J. Comput. Phys. **90**(1), 141–160 (1990)
4. Karni, S., Čanić, S.: Computations of slowly moving shocks. J. Comput. Phys. **136**(1), 132–139 (1997)
5. Xu, K., Hu, J.: Projection dynamics in godunov-type schemes. J. Comput. Phys. **142**(2), 412–427 (1998)
6. Pantano, C., Saurel, R., Schmitt, T.: An oscillation free shock-capturing method for compressible van der waals supercritical fluid flows. J. Comput. Phys. **335**, 780–811 (2017)
7. Arora, M., Roe, P.L.: On postshock oscillations due to shock capturing schemes in unsteady flows. J. Comput. Phys. **130**(1), 25–40 (1997)
8. Zaide, D., Roe, P.: Shock capturing anomalies and the jump conditions in one dimension. In: 20th AIAA Computational Fluid Dynamics Conference (2011)
9. Roe, P.L.: Approximate riemann solvers, parameter vectors, and difference schemes. J. Comput. Phys. **43**(2), 357–372 (1981)
10. Davis, S.F.: A simplified TVD finite difference scheme via artificial viscosity. SIAM J. Sci. Stat. Comput. **8**(1), 1–18 (1987)
11. Guardone, A., Vigevano, L.: Roe linearization for the van der waals gas. J. Comput. Phys. **175**(1), 50–78 (2002)
12. Randall J.L.: Finite volume methods for hyperbolic problems. Cambridge Texts in Applied Mathematics. Cambridge University Press (2002)

Numerical Modelling and Methods

Evaluating Differences Between the Computable and Filtered State Variables in Supercritical Carbon Dioxide

Benedikt Lea, Wojciech Sadowski, and Francesca di Mare

Abstract Supercritical CO_2-based power cycles require consideration of non-ideal thermodynamics and compressible turbulence. In present work, we study a decay of homogeneous isotropic turbulence with eddy shocklets using Direct Numerical Simulation. We consider both ideal gas and use Span and Wagner equation of state to model supercritical CO_2. The results are explicitly filtered, allowing for evaluation of a previously untested, yet universally accepted assumption, that the computable temperature and pressure (outputs of the equation of state using the available filtered quantities) are good approximations for the true filtered temperature and pressure. For the investigated thermodynamic state, the approximation does not result in a great error, however it does influence both the mean and the shape of PDF of both variables.

Keywords sCO_2 · Computable state variables · Compressible DNS · LES

1 Introduction

From a techno-economical point of view, supercritical CO_2-based power cycles represent a promising future energy systems technology [1], albeit a challenging one from the design perspective, as supercritical CO_2 (sCO_2) in turbomachines and heat exchangers is governed by non-ideal thermodynamics [2]. To benefit from the liquid-like behavior of sCO_2 in the compression step, the compressor inlet state has to be close to the vapor-liquid critical point, where potential condensation at the blade leading edge may occur [3]. The large heat transfer rates close to the Widom line,

B. Lea (✉) · W. Sadowski · F. di Mare
Ruhr University Bochum, Bochum, Germany
e-mail: benedikt.lea@rub.de

W. Sadowski
e-mail: wojciech.sadowski@rub.de

F. di Mare
e-mail: francesca.dimare@rub.de

A. Giauque et al. (eds.), *Proceedings of the 5th International Seminar on Non-Ideal Compressible Fluid Dynamics for Propulsion and Power*, ERCOFTAC Series 32,
https://doi.org/10.1007/978-3-031-92695-2_5

where the heat capacity is maximum, poses major challenges on materials and overall heat exchanger designs for durable operation [4]. Interactions between potential two-phase flow, the presence of broadband turbulence and compressibility effects further increase the complexity of the subject. To facilitate fundamental research into these phenomena, compressible Large Eddy Simulation (LES) seems to be the most promising method.

When Large Eddy Simulation is performed, a low-pass filtering operation is applied to the Navier–Stokes (NS) equations, resulting in a system expressed in terms of the filtered quantities [5], denoted here with the overbar $\overline{\cdot}$. The contributions from neglected subgrid-scales (SGS) must be modeled to close the system, with the accuracy of the overall method depending both on the numerical schemes used to discretize the equations and the quality of the turbulent closure. The low-pass filtering operator is defined as the convolution product of a filtering kernel $G(\boldsymbol{x}, \Delta)$ with the filtered variable, i.e., $\overline{\psi}(\boldsymbol{x}, t) = G(\boldsymbol{x}, \Delta) \star \psi(\boldsymbol{x}, t)$, where Δ is the characteristic size of the kernel, defining also the cut-off frequency of the filter. For variable-density flows, it is common practice to adopt a density-weighted filtering operator, resulting in a Favre filtering i.e., $\widetilde{\psi} = \overline{\rho\psi}/\overline{\rho}$.

To close the set of conservation laws governing fluid flows, thermal and caloric equations of state (EoS) are needed in order to model the interactions between the molecular and continua levels, and thereby, compute thermodynamic pressure p and temperature T from the vector of conserved variables. The application of Gibbs' phase rule to the single-phase pure gas flows in thermodynamic equilibrium states that thermophysical properties of such systems possess only two degrees of freedom. Typically, the density ρ and the internal energy e are chosen as independent thermodynamic state variables for density based algorithms. Thermal and caloric equation of state that fulfill this requirement must have the form $p = \mathrm{EoS_T}(\rho, e)$ and $T = \mathrm{EoS_C}(\rho, e)$ respectively.

LES equations are usually formed in such a way that a filtered density $\overline{\rho}$ and Favre filtered internal energy $\widetilde{e}$ are available, and therefore, the equations of state are evaluated with these parameters, resulting in *computable* temperature and pressure, $\check{T} = \mathrm{EoS_C}(\overline{\rho}, \widetilde{e})$ and $\check{p} = \mathrm{EoS_T}(\overline{\rho}, \widetilde{e})$. In case of a perfect gas, the computable values are equivalent to the filtered variables due to the linearity of the thermal and caloric equation of state:

$$
\begin{aligned}
\overline{p} &= \overline{\mathrm{EoS_T}(\rho, e)} = (\gamma - 1)\overline{\rho}\widetilde{e} = \mathrm{EoS_T}(\overline{\rho}, \widetilde{e}) = \check{p}, \\
\widetilde{T} &= \widetilde{\mathrm{EoS_C}(\rho, e)} = \frac{(\gamma - 1)}{R}\widetilde{e} = \mathrm{EoS_C}(\overline{\rho}, \widetilde{e}) = \check{T}.
\end{aligned} \tag{1}
$$

In the supercritical and dense vapor region, the real fluid behavior must be taken into account by non-linear equations of state, in which the use of $\check{T}$ and $\check{p}$ (or, in other words, assuming that $\check{p} - \overline{p} \approx 0$ and $\check{T} - \widetilde{T} \approx 0$) leads to new SGS modeling errors.

Recent works in the combustion community explored this issue by assessing the SGS contributions using ideal gas law [6] and cubic equations of state [7], whilst studying turbulent mixing in reacting flows. Media with high molecular complexity, such as Bethe-Zel'dovich-Thompson fluids, have also been considered in context of *a priori* investigation of common SGS models accuracy [8]. Non-classical gas dynamics in turbine cascades using virial equations of state has also been explored [9]. Recently, Sengupta et al. studied heat transfer mechanisms in a low-Mach channel flow at supercritical pressures [10]. However, they employed a simplified description of the thermophysical properties by using the Peng-Robinson equation. To the best of our knowledge, the SGS contributions from state-of-art reference equations for low molecular complexity fluids, such as CO_2, has not been a topic of extensive research. Particularly, the Span-Wagner reference equation [11] (hereafter referred to as SW EoS) has not been tested in this context.

In present work, we assess the behavior of the SGS-contributions in the state equation in the case of decaying homogeneous isotropic turbulence at high turbulent Mach number, for CO_2 under supercritical conditions as an ideal gas and modeled by SW EoS. A set of three direct numerical simulation are performed, considering ideal gas, CO_2 in the limit of an ideal gas and CO_2 close to the vapor-liquid critical point. The results of each simulation are explicitly filtered, at different resolutions appropriate for an LES computation, allowing for direct comparison of computable and filtered quantities and assessment of the error of the $\check{p} - \overline{p} \approx 0$ and $\check{T} - \widetilde{T} \approx 0$ assumptions.

2 Mathematical Model

The present work considers the three-dimensional Navier–Stokes equations in conservative form,

$$\partial_t \boldsymbol{U} + \nabla \cdot \boldsymbol{F}^a - \nabla \cdot \boldsymbol{F}^d = 0, \tag{2}$$

where $\boldsymbol{U}$ represents the vector of conserved variables $\boldsymbol{U} := (\rho, \rho\boldsymbol{u}, \rho E)$ comprised of density ρ, specific momentum $\rho\boldsymbol{u}$ and specific total energy density ρE, related to the specific static internal energy e as $E = e + \boldsymbol{u} \cdot \boldsymbol{u}/2$. The advective flux $\boldsymbol{F}^a$ is given as

$$\boldsymbol{F}^a := (\rho\boldsymbol{u}, \rho\boldsymbol{u}\boldsymbol{u} + p\boldsymbol{I}, \rho\boldsymbol{u}H), \tag{3}$$

where p represents thermodynamic pressure, $\boldsymbol{I}$ is the identity matrix ($I_{ij} = \delta_{ij}$) and $H = E + p/\rho$ is the specific total enthalpy. The diffusive flux $\boldsymbol{F}^d$ is given as

$$\boldsymbol{F}^d := (0, \boldsymbol{\tau}, \boldsymbol{\tau} \cdot \boldsymbol{u} - \boldsymbol{q}), \tag{4}$$

where $\boldsymbol{\tau} = \mu(\nabla\boldsymbol{u} + \nabla\boldsymbol{u}^T - 2/3\nabla \cdot \boldsymbol{u}\boldsymbol{I})$ is the viscous stress tensor and $\boldsymbol{q} = -\lambda\nabla T$ is the heat flux vector.

The thermophysical properties for CO_2 are calculated according to the state-of-art reference EoS by Span and Wagner [11]. The fundamental equation is expressed in the form of the dimensionless Helmholtz energy $\phi = A/RT$, which is a function of reduced density $\delta = \rho/\rho_c$ and inverse reduced temperature $\tau = T_c/T$. For the reduction, the critical density $\rho_c = 467.6\text{kg/m}^3$ and the critical temperature $T_c = 304.128\text{K}$ are used respectively. To model the ideal-gas behavior and the residual fluid behavior, the dimensionless Helmholtz energy is commonly split into two distinct terms $\phi = \phi^{\text{o}} + \phi^{\text{r}}$, where the superscript o indicates the ideal-gas part and r the residual part. The functional form consists of 42 highly non-linear terms, whereby the residual fluid behavior is solely described empirically, by optimizing and fitting the coefficients to experimental data. Since the Helmholtz energy is one form of a fundamental equation, all the thermodynamic properties of a pure substance, such as the internal energy can be obtained by combining its derivatives

$$e(\delta, \tau) = RT\tau(\phi_\tau^{\text{o}} + \phi_\tau^{\text{r}}). \tag{5}$$

Since the SW EoS is not directly expressed in terms of density and internal energy, an iterative procedure is required to determine the temperature. In the current work, we use the Newton-Rhapson method to find the roots of

$$f(T) := e - e(\rho, T)|_\rho. \tag{6}$$

Then, the pressure is given by $p = \rho RT(1 + \phi_\delta^{\text{o}})$. As we are mainly concerned with the influence of the SW EoS in this work, we assume the viscosity μ to follow a power-law

$$\mu/\mu_{ref} = (T/T_{\text{ref}})^{3/4}, \tag{7}$$

which only models the dependence of temperature and specify a constant Prandtl number of 0.71 for the thermal conductivity λ.

For the kernel to be adequate for the filtering of Navier–Stokes equations, it must fulfill properties of linearity and normalization. It is also assumed here that the kernel is homogenous and commutes with derivatives. In the present work, the results of the resolved computations are filtered with the three-dimensional top-hat filter [5], given as

$$G(\boldsymbol{x}, \Delta) = \begin{cases} \Delta^{-3}, & \text{if } |x_i| < \Delta/2 \,\&\, i = 1, 2, 3 \\ 0, & \text{otherwise.} \end{cases} \tag{8}$$

3 Decay of Homogeneous Isotropic Turbulence (DHIT)

We use an in-house GPU-accelerated finite difference solver intended for DNS of turbulent compressible flows using a hybrid flux formulation, which was presented and initially validated in [12]. Following the work done in [13, 14], we solve the NS equations in a periodic domain with a size $L = 2\pi$ discretized with a uniform mesh with 128^3 computational nodes. The flow is defined by a turbulent Mach number $\text{M}_t = \sqrt{\langle u_i u_i \rangle}/\langle c \rangle$ and Taylor-scale Reynolds number $\text{Re}_\lambda = \langle \rho \rangle\, u_{\text{rms}} \lambda / \langle \mu \rangle$, based on the speed of sound c, root-mean square of velocity fluctuations $u_{\text{rms}} = \sqrt{\langle u_i u_i \rangle / 3}$ and Taylor microscale $\lambda = \langle u_1^2 \rangle / \langle (\partial_1 u_1)^2 \rangle$. In all of the above, $\langle \cdot \rangle$ denotes an average in whole domain. All of the simulations were run with $\text{M}_{t,0} = 0.6$ and $\text{Re}_{\lambda,0} = 100$. The computations are performed for time of $t/\tau = 4$, where the eddy turn-over time is defined as $\tau = \lambda_0 / u_{\text{rms},0}$.

In the flow under investigation, the Kolmogorov length scale η can be estimated *a priori* as [15]

$$\eta = \lambda 15^{-1/4} \text{Re}_\lambda^{-1/2}, \tag{9}$$

which leads to the mesh resolution of approximately $\Delta_{\text{DNS}} \approx 1.9\eta$ of the performed DNS. For an appropriately resolved LES, around 80% of turbulent kinetic energy has to be resolved, which corresponds to the mesh resolution $\Delta \approx 25\eta_0$ [16]. Hence, to study the errors occurring by the evaluation of $\widetilde{T}$ and $\overline{p}$, we adopted three filter sizes with $\Delta/\Delta_{\text{DNS}} = 3, 5, 9$, which, based on Eq. (9) corresponds to a highly-resolved LES with $\Delta/\eta_0 \approx 5.8, 9.7, 17.4$.

As in [14], the initial random and solenoidal velocity field is generated using the method by Rogallo [17], from Passot-Pouquet [18] spectrum of energy given as $E(k) \propto k^4 \exp(-2(k/k_0)^2)$ with the integral of the spectrum equal to $3u_{\text{rms}}^2/2$ and $k_0 = 2/\lambda_0 = 4$ determining the most energetic wave number. Initial conditions for density and internal energy are specified as constant values.

Three separate simulations of DHIT have been performed. To facilitate comparison with results from [14], first an ideal gas flow was considered ($\gamma = 1.4$, $R = 287$). In the text, this computation is referred to with **IG**. To further validate the implementation of the equation of state, DHIT in CO_2 in the limit of an ideal gas (compressibility factor $Z = 1$) was simulated. Results of this simulation will be referred to as **CO_2I**. Finally, a computation with non-ideal gas behavior close to critical point (named **CO_2NI**) was performed ($Z = 0.53$). For each simulation, the initial conditions and reference values of viscosity and temperature used to match the Reynolds number are given in Table 1.

Table 1 Initial and reference conditions for each simulation

	p_0 [Pa]	ρ_0[kg/m^3]	$u_{\text{rms},0}$[m/s]	μ_{ref}[Ns/m^2]	T_{ref} [K]
IG	1×10^5	1	129.6	0.64807	348.43
CO_2I	1×10^5	1	121.2	0.60596	529.61
CO_2NI	77.5×10^5	246.49	73.06	90.045	90.045

4 Results

4.1 Validation in the Ideal Gas Limit

As the simulation time advances, the vortical structures from the initial conditions break down and decay. Figure 1 shows the difference between the initial condition and the turbulence at $t/\tau = 4$. The evolution of kinetic energy, enstrophy $\langle \omega_i \omega_i \rangle /2$, where $\boldsymbol{\omega} = \nabla \times \boldsymbol{u}$ is the vorticity, temperature variance and dilatation over the whole computation time of 4τ for the cases **IG** and **CO_2I** is presented in Fig. 2. Overall, the ideal gas data matches the reference solution to a sufficient degree, with small differences visible in the enstrophy and dilatation. The reference statistics presented in [14] have been filtered to a resolution of 64^3, therefore a non-perfect match is expected.

Shifting focus to the **CO_2I** case, we can observe that the results obtained using the ideal gas state equations are reproduced almost exactly (again with small differences in dilatation) with the SW EoS operating with $Z = 1$, thereby further validating the implementation. As expected, the values of all parameters computed with filtered velocity and temperature are consistently lower than the ones from unfiltered DNS. The appropriateness of the choice of the filter size can be assessed by inspecting the

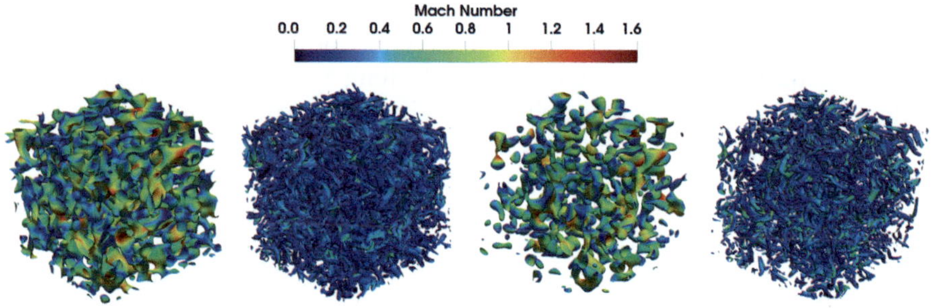

Fig. 1 Vortical structures visualization using isosurfaces of Q-criterion with $Q = 1 \times 10^5$ colored by Mach number. From left to right: the initial field of **CO_2I** case, flow field at $t/\tau = 4$ for **CO_2I**, the initial field of **CO_2NI** case and flow field at $t/\tau = 4$ for **CO_2NI** simulation

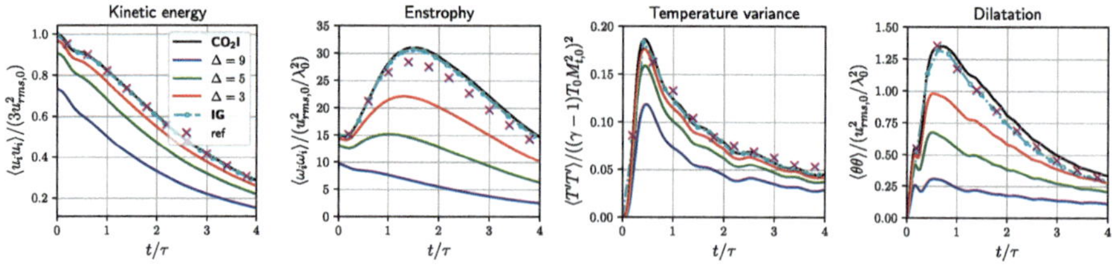

Fig. 2 Time series of kinetic energy, enstrophy, temperature variance and dilatation, from the cases **IG** and **CO_2I**, compared to [14]. Additionally, same parameters computed using filtered quantities are presented

plots of filtered and resolved turbulent kinetic energy in Fig. 2, and evaluating the fraction of resolved turbulent kinetic energy *a posteriori*. Filtering at $\Delta/\Delta_{\mathrm{DNS}} = 3$ and $\Delta/\Delta_{\mathrm{DNS}} = 5$ allows for resolution of more than 80% of kinetic energy. The largest filter does not meet this requirement and, contrary to the results of the *a priori* estimate, can be considered representative of under-resolved LES. This difference could be attributed to the relative inaccuracy of the estimation of η due to relatively low Reynolds number in the performed simulation. However, it is important to note that the *a priori* estimation of grid resolution for LES is difficult and usually only *a posteriori* estimation of the quality of LES can be considered accurate.

To ensure that the numerical schemes are able to resolve turbulence accurately assuming the non-ideal equation of state, the spectra of turbulent kinetic energy E_k and vorticity E_ω has been computed for $t/\tau = 4$ (Fig. 3). The results agree with the data reported in [14]. Furthermore, both spectra have been computed using $\widetilde{\boldsymbol{u}}$ for each of the studied filter sizes. As expected, due to the non-monotonicity of the transfer function of the top-hat filter, both E_k and E_ω exhibit oscillations in the range of high wave numbers.

Furthermore, the property given by Eq. (1) was tested. The differences $\overline{p} - \check{p}$ and $\widetilde{T} - \check{T}$ were directly computed for each time step (not plotted here). In order to assess the significance of those differences depending on the EoS used, we opted to normalize them by the standard deviation of the SGS components of p and T.

For each filter size, the standard deviation was computed as $\sigma(\psi_{\mathrm{sgs}}) = \sqrt{\langle\psi\psi\rangle - \langle\overline{\psi}\,\overline{\psi}\rangle}$. Both, $\sigma(p_{\mathrm{sgs}})$ and $\sigma(T_{\mathrm{sgs}})$ represent a scale of variation of the components of pressure and temperature that are neglected in implicit LES, therefore, this normalization allows for comparing the error resulting from the use of computable values to the error already typically committed in the context of LES. Furthermore, in order to fully assess the importance of these terms, their gradients should be evaluated in the form as present in the filtered governing equations [19],

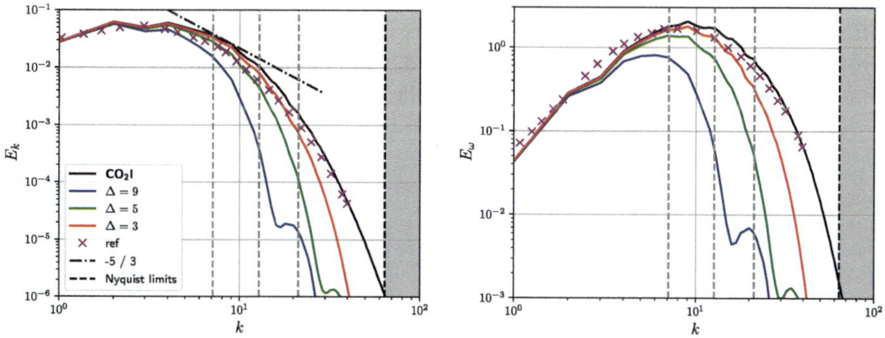

Fig. 3 Turbulent kinetic energy (left) and vorticity (right) spectra computed from $\mathbf{CO_2I}$ simulation at $t/\tau = 4$ (along with spectra computed from the filtered quantities), compared to the reference results from [14]. The vertical dashed lines correspond to the cut-off frequencies of the filters and the Nyqust limit of the DNS mesh

and then compared to each other. This is however, outside the scope of the current manuscript and will be the subject of further investigations.

For each time step and filter size, the values of the error remained below $0.005\sigma(T_{sgs})$ and $0.001\sigma(p_{sgs})$ for the temperature and pressure respectively. The standard deviations σ and skewness of each distribution (both filtered and computable) were also compared showing also no significant differences.

4.2 Simulation at the Non-Ideal Thermodynamic State

Figure 1 also presents the initial condition of $\mathbf{CO_2NI}$ simulation and the evolved turbulence at $t/\tau = 4$. Simulation $\mathbf{CO_2NI}$ is started with different set of initial conditions, which is reflected in the shape of isocontours of Q. Time series of turbulent kinetic energy, enstrophy and temperature variance presented in Fig. 5 remain very close to the results obtained from the $\mathbf{CO_2I}$ simulation. Interestingly, the non-ideal thermodynamic state seems to decrease the dilatation of velocity w.r.t. the ideal gas conditions. Nevertheless, considering these results, and the spectra E_k and E_ω (not shown here) which agree with the ones computed in $\mathbf{CO_2I}$ case, the simulation $\mathbf{CO_2NI}$ can be deemed sufficiently validated (Fig. 4).

The mean of the $\widetilde{T} - \check{T}$ difference is visible in Fig. 5. For the smallest filter size, the differences remain larger than $0.02\sigma(T_{sgs})$ for the majority of the simulation time. The values of the error rises with the increasing filter size, which is an expected behavior. Similar situation can be visible in the error of pressure computation, however, the values here are much smaller, i.e., for the largest filter size, the error remains below $0.025\sigma(p_{sgs})$ for the whole computation time outside of the initial peak.

The difference in the distributions of computable and filtered variables is captured by comparing differences of standard deviations and skewnesses of each PDF. Interestingly, the difference in standard deviation of temperature is mainly negative, whereas for pressure it remains positive for the simulation after the initial peak. The difference in standard deviations of pressure is small in magnitude. The magnitude of errors for temperature and pressure standard deviations seem to be the largest for the finest filter width. The results indicate, that the standard deviation SGS fluctuations

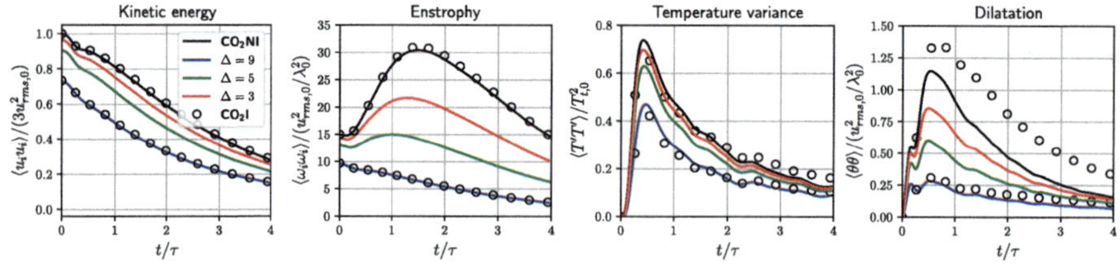

Fig. 4 Time series of kinetic energy, enstrophy, temperature variance and dilatation, from $\mathbf{CO_2NI}$ simulation compared to $\mathbf{CO_2I}$ results. Time series evaluated using filtered quantities are also reported

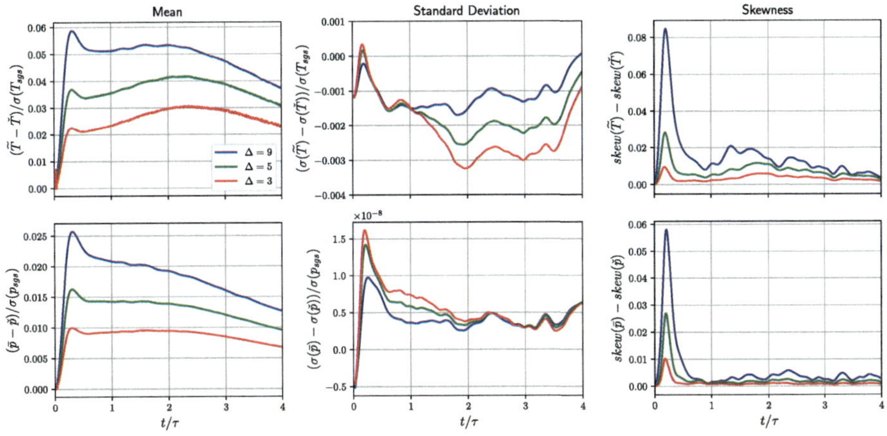

Fig. 5 case, from left to right: the mean difference between the computable and filtered p and T normalized by σ of corresponding SGS fluctuations, difference of standard deviations of PDFs of both quantities normalized in the same way and the differences between skewness of both distributions

tends to approach zero much faster in the limit $\langle\overline{\psi}\,\overline{\psi}\rangle \rightarrow \langle\psi\,\psi\rangle$, than the numerator. To verify this hypothesis, it would be interesting to investigate forced HIT and compare statistically steady quantities. After the initial peak in skewness, the difference decreases over the decay and approaches a similar value for all considered filter sizes.

5 Conclusions and Outlook

The objective of the present work is to evaluate the difference between the computable pressure and temperature (values computed from the equation of state using filtered density and internal energy) and the true filtered distributions of the variables, thereby assessing the approximation usually adopted in Large Eddy Simulation of compressible flows of non-ideal gases. To do so, three Direct Numerical Simulations have been performed using highly efficient GPU implementation of high-order finite difference method, results of which has been filtered to LES resolution. Span and Wagner equation of state has been used to model the thermophysical properties of CO_2.

For the non-ideal thermodynamic state, results indicate that the magnitudes of the differences between computable and filtered quantities do not represent significant errors when compared to the standard deviation of SGS fluctuations, a measure of the scales typically neglected in LES. On the other hand, the analysis shows, that the computable parameters have slightly different PDFs, as indicated by differences in standard deviation and skewness. The differences increase with increased filter size, indicating another potential source of error present in under-resolved LES.

The present investigation offers only a preliminary insight into the topic, and there are several questions that we plan to address in further studies. First and foremost, all the terms in the momentum and energy equations, including the error terms containing the differences between the computable and filtered quantities, should be evaluated and compared with each other, to assess the importance of the terms in the context of each equation. Additionally, in the present case, we evaluate the error based on accurate DNS results, therefore, the inaccurately computed state variables do not have the chance to influence the evolution of conservation equations, which could increase the error in real LES. This could be tested by comparing LES with DNS results or by performing an explicit LES to truly isolate the influence of the studied approximation. Another important extension of the current study would be to perform the same analysis in forced isotropic turbulence, thereby removing the influence of initial conditions and enabling gathering of statistically time-invariant parameters.

Acknowledgements The investigations were conducted as part of the joint research program DigITecT in the frame of AG Turbo. The work was supported by the Federal Ministry of Economic Affairs and Climate Action (BMWK) as per resolution of the German Federal Parliament under grant number 03EE5117H. Benedikt Lea gratefully acknowledges AG Turbo for the support and permission to publish this paper. Wojciech Sadowski would like to gratefully acknowledge the funding by the Deutsche Forschungsgemeinschaft (DFG, German Research Foundation) through SFB/TRR287, Project Number 422037413. The responsibility for the content lies solely with its authors.

References

1. Angelino, G.: Carbon dioxide condensation cycles for power production. J. Eng. Power **90**(3), 287–295 (1968)
2. Alberto, G., Piero, C., Matteo, P., Andrea, S.: Nonideal compressible fluid dynamics of dense vapors and supercritical fluids. Ann. Rev. Fluid Mech. **56**, 241–269 (2024)
3. Lettieri, C., Paxson, D., Spakovszky, Z., Bryanston-Cross, P.: Characterization of nonequilibrium condensation of supercritical carbon dioxide in a de Laval nozzle. J. Eng. Gas Turb. Power **140**(4), 041701 (2018)
4. Klaus, B., Richard, D., Peter, F.: Fundamentals and applications of supercritical carbon dioxide (Sco2) based power cycles, 1st edn. Elsevier, Waltham, MA (2017)
5. Garnier, E., Adams, N., Sagaut, P.: Large Eddy simulation for compressible flows. Scientific computation. Springer, Netherlands (2009)
6. Ribert, G., Domingo, P., Vervisch, L.: Analysis of sub-grid scale modeling of the ideal-gas equation of state in hydrogen-oxygen premixed flames. Proceed. Combust. Inst. **37**(2), 2345–2351 (2019)
7. Unnikrishnan, U., Oefelein, J.C., Yang, V.: Subgrid modeling of the filtered equation of state with application to real-fluid turbulent mixing at supercritical pressures. Phys. Fluids **34**(6), 065112 (2022)
8. Giauque, A., Giguet, C., Vadrot, A., Corre, C.: A priori test of Large-Eddy simulation models for the Sub-Grid Scale turbulent stress tensor in perfect and transcritical compressible real gas Homogeneous Isotropic Turbulence. Comput. Fluids **268**, 106091 (2024)
9. Hoarau, J.-C., Cinnella, P., Gloerfelt, X.: Large Eddy simulations of strongly non-ideal compressible flows through a transonic cascade. Energies **14**(3), 772 (2021)

10. Sengupta, U., Nemati, H., Boersma, B.J., Pecnik, R.: Fully compressible low-mach number simulations of carbon-dioxide at supercritical pressures and trans-critical temperatures. Flow, Turb. Combust. **99**(3), 909–931 (2017)
11. Span, R., Wagner, W.: A new equation of state for carbon dioxide covering the fluid region from the triple-point temperature to 1100 K at pressures up to 800 MPa. J. Phys. Chem. Ref. Data **25**(6), 1509–1596 (1996)
12. Pascal, P., di Francesca, M.: Towards direct numerical simulations of shock-turbulence interaction in real gas flows on GPUs: Initial validation. In: di Francesca, M., Andrea, S., Matteo, P. (eds.) Non-Ideal Compressible Fluid Dynamics for Propulsion and Power, pp. 59–75. Springer International Publishing, Cham (2020)
13. Lee, S., Lele, S.K., Moin, P.: Eddy shocklets in decaying compressible turbulence. Phys. Fluids A: Fluid Dyn. **3**(4), 657–664 (1991)
14. Eric, J., Johan, L., Ankit, V.B., William, H.C., Parviz, M., Britton, J.O., Pradeep, S.R., Santhosh, K.S., Björn, S., Yee, H.C., Xiaolin, Z., Sanjiva, K.L.: Assessment of high-resolution methods for numerical simulations of compressible turbulence with shock waves. J. Comput. Phys. **229**(4), 1213–1237 (2010)
15. Tennekes, H., John, L.L.: A first course in turbulence. MIT Press (1972)
16. Celik, I.B., Cehreli, Z.N., Yavuz, I.: Index of resolution quality for large eddy simulations. J. Fluids Eng. **127**(5), 949–958 (2005)
17. Rogallo, S.: Numerical experiments in homogeneous turbulence. NASA Technical memorandum 81315, NASA (1981)
18. Passot, T., Pouquet, A.: Numerical simulation of compressible homogeneous flows in the turbulent regime. J. Fluid Mech. **181**(1), 441 (1987)
19. Giauque, A., Corre, C., Menghetti, M.: Direct numerical simulations of homogeneous isotropic turbulence in a dense gas. J. Phys. Conf. Ser. **821**(1), 012017 (2017)

Towards a Thermodynamic Validation Map for Non-Ideal Compressible Flows

Diogo S. C. Fernandes, Matteo Majer, and Adam J. Head

Abstract Computational models capable of simulating non-ideal compressible flows are not yet sufficiently validated against measurement data. Furthermore, the wide variety of thermodynamic models available together with the large operational ranges make it difficult to correctly construct the simulation model with a sufficient balance between accuracy and computational speed. This research assessed the impact of the choice of various thermodynamic models on the accuracy of flow solvers, under a wide variety of operating conditions. This will allow for the generation of a thermodynamic map from which validation experiments can be designed. To that end, the results provided by polytropic (PPR) and non-polytropic (NPPR) Peng-Robinson EoS-based simulations in the SU2 flow solver were compared to those by a multi-parameter Helmholtz equation of state (HEOS), on a grid of 144 operating conditions, over the ORCHID nozzle domain. The pressure and Mach number responses were compared, as these can be measured experimentally, and previously obtained experimental uncertainties (assumed to be constant) were used as a distinguishing criterion. Mach number comparisons didn't allow to distinguish between the models because of the high measurement uncertainty. In terms of pressure response, deviations were found to be negligible (lower than experimental uncertainties) below around $\frac{p}{p_{\text{crit}}} = 0.31$. Further comparing across the whole grid, deviations increased to over 10 times experimental uncertainty values at supercritical conditions, increasing with decreasing compressibility factor. Additionally, Peng-Robinson EoS simulations were five times faster than HEOS, emphasising why one would choose them if the deviations are acceptably low. It is recommended to design validation experiments using the resulting map, and to further investigate the physical causes of these deviations.

D. S. C. Fernandes (✉) · M. Majer · A. J. Head
Propulsion & Power, Delft University of Technology, Delft, The Netherlands
e-mail: d.sousacabralfernandes@student.tudelft.nl

M. Majer
e-mail: m.majer@tudelft.nl

A. J. Head
e-mail: a.j.head@tudelft.nl

A. Giauque et al. (eds.), *Proceedings of the 5th International Seminar on Non-Ideal Compressible Fluid Dynamics for Propulsion and Power*, ERCOFTAC Series 32,
https://doi.org/10.1007/978-3-031-92695-2_6

Keywords ORC · Hexamethyldisiloxane (MM) · Peng-Robinson · Helmholtz equation of state · SU2

1 Introduction

As society transitions to sustainable energy production, adapting and improving technologies is necessary to meet increasing energy demands and reduce emission outputs [1]. One such technology are Organic Rankine Cycle (ORC) systems, which are quite attractive given their capability to convert low to medium temperature heat—notably waste heat from other systems—into power in an efficient and cost-effective manner [2]. Such systems make use of an organic substance as a working fluid, which has a number of advantages compared to the water used in traditional Rankine cycles when it comes to low temperature heat applications [3].

Small-scale ORC technology has not yet, however, reached a commercially viable stage, on account of the challenges related to its design and optimisation [3]. These are due to the properties of the organic fluids used, which give rise to highly non-ideal supersonic flows in the turbine [4]. Modelling these complex flows thus requires equally complex thermodynamic models, in order to close the applicable conservation equations [5], and it is often not clear which model should be used [6]. Such models include the Peng-Robinson cubic equation of state (EoS) [7, 8], which can be formulated in a polytropic (PPR) version, and a non-polytropic (NPPR) one [9] depending on the chosen caloric model. Alternatively, more complex models have been proposed, such as multi-parameter Helmholtz equations of state (HEOS) [10, 11].

Computational fluid dynamics (CFD) programs—such as the SU2 fluid flow solver [12]—have become increasingly accurate over the past decades, and have been used in Non-Ideal Compressible Fluid Dynamics (NICFD) applications [13]. A number of studies have worked towards validation of CFD solvers, and notably SU2, for NICFD flows [14–17], using experimental data from purpose-built test rigs, such as the Organic Rankine Cycle Hybrid Integrated Device (ORCHID) at TU Delft [6], and the Test-Rig for Organic VApours (TROVA) at Politecnico di Milano [18]. These solvers are not yet fully validated over a wide range of operating conditions, however. One aspect in which further research is required is related to the choice of the different thermodynamic models used by the software, particularly for off-design operating conditions of the test rigs used to generate validation data.

This paper aims to assess the impact of the choice of thermodynamic model and influence of the operating conditions on the predictive capabilities of CFD flow solvers. Several thermodynamic models are thus compared for a wide range of operating conditions. The computational performance of the models is also investigated, which one can use as an additional criterion for selecting the appropriate equation.

To this end, 144 simulations of the ORCHID planar nozzle [19] flow in off-design conditions were carried out using the SU2 flow solver using the polytropic Peng-Robinson, non-polytropic Peng-Robinson, and multi-parameter Helmholtz equations of state (implemented via CoolProp [20] in SU2).[1] The simulations were rusn for a variety of operating conditions in the $p - T$ thermodynamic plane, with the organic working fluid hexamethyldisiloxane (MM). To assess their accuracy, the resulting distributions of centerline Mach number and surface static pressure were compared. Experimental uncertaintines obtained from past test campaigns at the ORCHID [19] were notably used to determine the significance of the differences between models.

2 Methodology

In order to compare the three different thermodynamic models, a grid of upstream operating conditions that is, the pressure and temperature upstream of the nozzle was set up, and simulations were run with all these conditions. Figure 1 shows the grid of 144 points in the $p - T$ plane used in the simulations.

This grid is not uniform, and was obtained by starting with a sparse grid, and refining it as necessary throughout the study. This most notably involved refining near the area of the binary test threshold and in the supercritical area (see Fig. 4). The bounds of the grid were chosen as follows: $p = 2.5$ MPa is the maximum operating pressure of the ORCHID, $T = 493$ K and $T = 573$ K are the minimum and maximum temperatures considered that also correspond to the limits of accuracy of the HEOS method as implemented in CoolProp—with 573 K also being the upper thermal stability limit of MM [21, 22]—and $p = 0.4$ MPa is a low enough upstream pressure that, at high temperatures, allows for ideal gas expansion. Moreover, the grid excludes conditions where isentropic expansion would lead to two-phase flow, as this cannot currently be simulated.

Table 1 lists the remaining simulation parameters which were kept consistent between all simulations.

The mesh size was chosen based on previous mesh convergence studies [16]. Additionally, to ensure similar expansions (thermodynamically) through the nozzle throughout the design space investigated, the density ratio across the nozzle was kept constant and equal to $\frac{\rho_{in}}{\rho_{out}} = 14.2$, which corresponds to that at the design point of the nozzle. This was used to compute the nozzle density at the outlet, and consequently the back pressure, assuming an isentropic process. The specific heat ratio γ was specified as the average of the ideal specific heat ratio between the input and output of the nozzle [23].

HEOS was used as the reference model since it is expected to be the most accurate [11, 24], and then the deviations arising from the other models were compared. This comparison involved two parts. First, a binary test established whether the deviations were significant or not for each operating condition, defining a region where all models may be used without needing to consider their numerical accuracy. Second,

[1] The implementations of PPR and HEOS are available in the `develop` branch of SU2, while the NPPR implementation is available in the `feature_NPPR_GAS` branch.

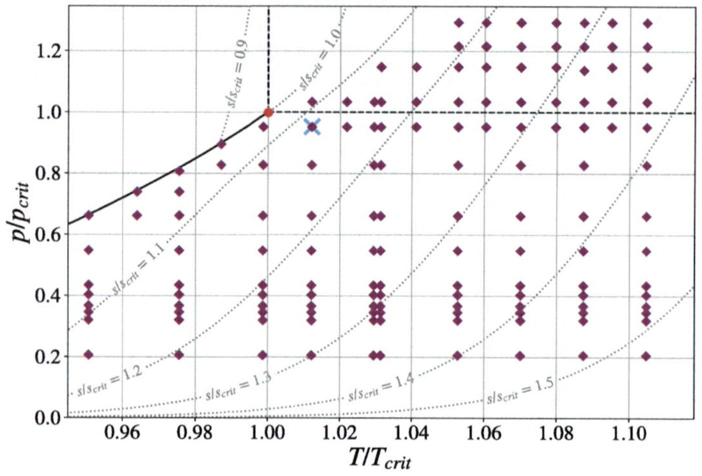

Fig. 1 Grid of operating conditions (⬩) considered, in the $p - T$ plane. Also represented are the design point (×) , saturation dome (—) and isentropes (……)

Table 1 Simulation parameters

Equations	Euler		
Mesh elements	67000		
Convergence order	−10		
Working fluid	MM (hexamethyldisiloxane)		
p_{crit}	1.9311	MPa	[11]
T_{crit}	518.7	K	[11]
s_{crit}	0.83384	kJ/kgK	[11]

an actual quantification of the deviations depending on the condition led to a heat map representing the varying accuracy of the models across the entire space of operating conditions.

The first comparison method was a binary test, that involved comparing the deviations with experimental uncertainties. Previously obtained experimental uncertainties at the design point [19] for the static pressure and Mach number were used, as these are the quantities that can be measured experimentally. These uncertainties were assumed to be constant regardless of operating conditions, as they are mostly due to instrument uncertainties. The deviations were then assessed based on whether they were smaller or greater than the experimental uncertainties, thus establishing whether or not they are significant. Figure 2 shows an example of how this procedure is conducted, for the ORCHID design point ($\frac{T}{T_{\text{crit}}} = 1.012$ and $\frac{p}{p_{\text{crit}}} = 0.9528$). Figure 2a shows that at this operating condition, the pressure deviations are lower than experimental uncertainties in the foremost part of the nozzle, but exceed them

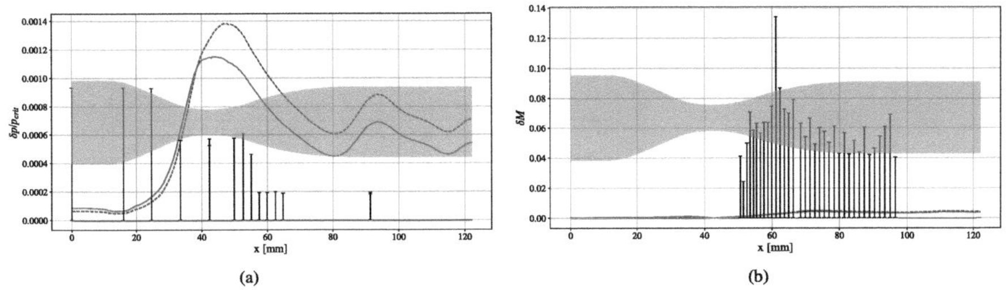

Fig. 2 Graph of deviations between the various thermodynamic models along the nozzle, at the design point ($\frac{T}{T_{\text{crit}}} = 1.012$ and $\frac{p}{p_{\text{crit}}} = 0.9528$), with experimental uncertainties (I) overlaid. HEOS (__) is taken as the baseline, and PPR (- - -) and NPPR (__) are compared to it. The nozzle profile is also drawn in the background. **a** Pressure deviations ($\delta p/p_{\text{crit}}$). **b** Mach number deviations (δM)

near the throat and in the diverging part of the nozzle (for both the PPR and the NPPR cases). This condition therefore fails the binary test with regards to pressure deviations. Figure 2b, however, shows that this condition passes the binary test in terms of Mach number, mostly because the Mach number uncertainties are very large. In fact, they are so large that at all operating conditions, the responses had deviations significantly smaller than experimental uncertainties. This implies that for this validation process, experimental Mach number results are not as useful as other quantities, namely static pressure.

The second method involved creating a heat map based on a single comparison metric. This allows for a straightforward overview of the effect of operating conditions. To ensure an equitable comparison, it was chosen to use the pressure deviation between the models at a single, constant location. This was the mean location of the maximum deviation, 46.6 mm for PPR and 45.9 mm for NPPR, both shortly past the throat location (42.4 mm [19]). The location of this maximum deviation varies marginally with operating condition, with standard deviation $\sigma = 1.85$ mm for PPR and $\sigma = 1.79$ mm for NPPR, under 2% of the nozzle length. This ensures that the highest deviations are always taken into account.

3 Results

Figure 3 shows the results of the binary test. The static pressure was chosen as the comparison metric due to its significantly higher (experimental) accuracy when compared to the Mach number measurements. From the figure, one can observe that the upstream pressure has the largest effect on the results, with deviations at operating pressures below $\frac{p}{p_{\text{crit}}} = 0.31$ being fully contained within the experimental pressure uncertainty. It can be further seen that at these low pressures, the PPR and NPPR implementations had very similar behaviour. Only at one condition ($\frac{p}{p_{\text{crit}}} = 0.3625$

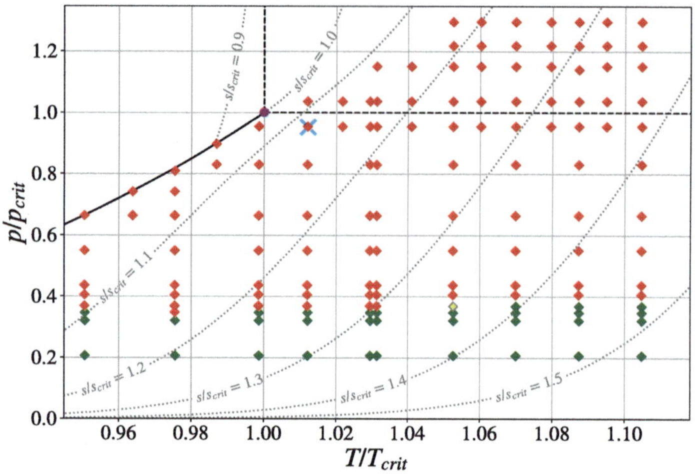

Fig. 3 Grid of simulated operating conditions ($p - T$ plane), coloured according to whether model pressure deviations are lower (◆) or higher (◆) than experimental uncertainties. One point (◆), $\frac{p}{p_{\text{crit}}} = 0.3625$ and $\frac{T}{T_{\text{crit}}} = 1.052$, has deviations lower than experimental uncertainties with PPR, but higher with NPPR. Also represented are the design point (×), saturation dome (—) and isentropes (........)

and $\frac{T}{T_{\text{crit}}} = 1.052$), PPR had lower deviations than experimental uncertainties, while NPPR had higher ones, though both were very close to the threshold.

Heat maps of the maximum wall static pressure deviation between cubic models and the reference HEOS model were generated in the thermodynamic plane $p - T$. Figure 4 presents these heat maps. The plots confirm the previous observation that the operating pressure has an overall bigger effect on the deviations than the operating temperature, though this is only the case at subcritical conditions, and far from the saturation dome. Furthermore, one can observe that the deviations are comparatively low at subcritical conditions, but rapidly increase at supercritical conditions. In the supercritical region, pressure deviations increase inversely proportional to the compressibility factor Z. The highest deviations observed were $\frac{\delta p}{p_{\text{crit}}} = 0.0057$ with PPR and $\frac{\delta p}{p_{\text{crit}}} = 0.0050$ with NPPR. This shows that, even at supercritical conditions, NPPR and PPR continue to perform very similarly, even if NPPR results in slightly higher accuracy. In both cases, the highest deviation was observed at the lowest temperature simulated for the highest pressure considered: $\frac{T}{T_{\text{crit}}} = 1.053$ and $\frac{p}{p_{\text{crit}}} = 1.295$. These are therefore particularly interesting conditions to use for validation experiments.

Finally, it is important to note the difference in computational performance between the three models. While the time taken for a simulation varied greatly depending on operating condition, the variation was equal for all models, which maintained a consistent average time per iteration. The variation is instead explained

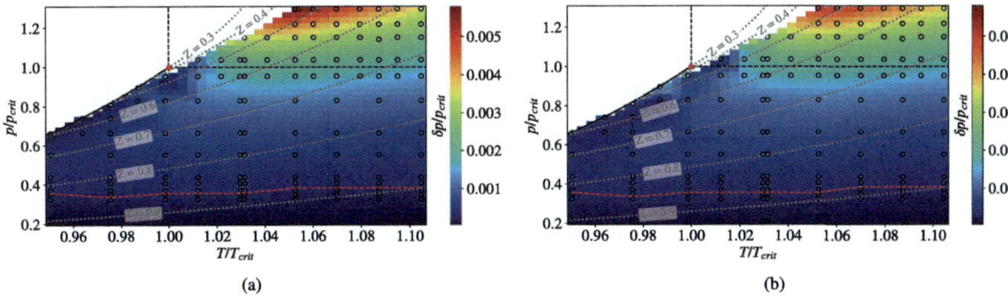

Fig. 4 Heat maps of the pressure deviation ($\frac{\delta p}{p_{\text{crit}}}$) at average maximum location between the cubic equations of state and HEOS, depending on the chosen operating condition ($p - T$ plane). Also represented are the input points (o), saturation dome (—), threshold from the experimental uncertainty-based binary test (---) and contour lines of the compressibility factor Z (......). **a** Heat map for PPR, at $x = 46.6$ mm. **b** Heat map for NPPR, at $x = 45.9$ mm

Table 2 Time per iteration depending on thermodynamic model

Model	Avg. Time per Iteration	
	Absolute [s]	Relative to PPR
PPR	0.204	1.00
NPPR	0.225	1.10
HEOS	1.15	5.64

by the number of iterations, which varied depending on operating conditions, but was very similar for all three models. The main difference between the models was therefore in the average time taken per iteration, which is presented in Table 2. It can be seen that the PPR and NPPR implementations have very similar performance (though NPPR is around 10% slower), and are both around 5 times faster than the HEOS implementation. This highlights the importance of comparing the accuracy between the models, as significant computational time can be gained by switching to a Peng-Robinson EoS, when these do not deviate too much from the HEOS baseline.

4 Conclusions and Future Work

This research evaluated the impact of the thermodynamic model and operating conditions on the accuracy of the flow solver available for the simulation of non-ideal compressible flows, applied to a planar de Laval nozzle, at off-design conditions. The evaluation was performed by simulating the fluid flow of MM (hexamethyldisiloxane) using cubic equations of state (EoS)—polytropic Peng-Robinson (PPR) and non-polytropic Peng-Robinson (NPPR)—and a multi-parameter Helmholtz equation of state (HEOS), over a mesh representative of the ORCHID nozzle profile, at a variety of operating conditions. The HEOS results were taken as a baseline and those of PPR and NPPR were compared to it. The resulting deviations between the models were first compared to previously obtained experimental uncertainties, to determine at which conditions deviations are negligible. Then, a heat map of the deviations at a specific location (the average location of maximum deviation, which has a very tight distribution) was generated, to more finely assess the evolution of deviations with operating conditions.

Several conclusions can be drawn from this study. First, it was found that current centerline Mach number measurements are not a useful validation metric as its measurements have very high experimental uncertainties. Then, it was found that the cubic equation of state (PPR and NPPR) implementations are, on average, 5 times faster than the HEOS implementation, highlighting the advantage of selecting them when deviations are sufficiently low for the study being conducted. This is always the case at lower operating pressures, with deviations found to be negligible (lower than experimental uncertainties) at operating pressures below approximately $\frac{p}{p_{\text{crit}}} = 0.31$. Furthermore, it was found that deviations remain relatively low at subcritical conditions, but are much larger at supercritical conditions, rapidly increasing with decreasing compressibility factor Z. Based on this, the cubic EoS are not appropriate if high accuracy at supercritical operating conditions is desired.

Future research should focus on designing and conducting new experiments using the results of this research to provide comprehensive validation data. Furthermore, other thermodynamic models, such as lookup tables, should be investigated for potential improvements in accuracy and computational efficiency. These investigations should also be combined with uncertainty quantification studies to assess the impact of model uncertainties on simulation results. More complex test cases, such as linear cascades, can also be considered in order to evaluate model performance under a wider range of conditions. Additionally, further exploring the reasons behind observed deviations and their change with varying operating conditions, would be beneficial to the development of best practices for simulating NICFD fluid flows. Addressing these areas can significantly improve the predictive capabilities and computational performance of CFD tools like SU2, thereby contributing to the design and integration of ORC systems in sustainable energy applications.

Acknowledgements The authors would like to thank Dr. Matteo Pini for his advice and feedback in the conduction of this research.

References

1. International Energy Agency: World Energy Outlook 2022. Technical report, IEA, Paris, France (2022)
2. Tchanche, B.F., Loonis, P., Pétrissans, M., Ramenah, H.: Organic Rankine Cycle Systems : Principles, opportunities and challenges. In: 2013 25th International Conference on Microelectronics (ICM), pp. 1–4, Beirut, Lebanon. IEEE (2013)

3. Colonna, P., Casati, E., Trapp, C., Mathijssen, T., Larjola, J., Turunen-Saaresti, T., Uusitalo, A.: Organic rankine cycle power systems: from the concept to current technology, applications, and an outlook to the future. J. Eng. Gas Turbines and Power **137**(10), 100801 (2015)
4. Spinelli, A., Cammi, G., Gallarini, S., Zocca, M., Cozzi, F., Gaetani, P., Dossena, V., Guardone, A.: Experimental evidence of non-ideal compressible effects in expanding flow of a high molecular complexity vapor. Exp. Fluids **59**(8), 126 (2018)
5. Reynolds, W.C., Colonna, P.: Thermodynamics: Fundamentals and Engineering Applications, 1st edn. Cambridge University Press, Cambridge, United Kingdom (2018)
6. Head, A.J.: Novel Experiments for the Investigation of Non-Ideal Compressible Fluid Dynamics: The ORCHID and First Results of Optical Measurements. Ph.D. thesis, Delft University of Technology (2021)
7. Stryjek, R., Vera, J.H.: PRSV: an improved Peng-Robinson equation of state for pure compounds and mixtures. Can. J. Chem. Eng. **64**(2), 323–333 (1986)
8. Smith, J.M., Van Ness, H.C., Abbott, M.M., Swihart, M.T.: Introduction to Chemical Engineering Thermodynamics, McGraw-Hill Education (2022)
9. Monjas, B.F.: Validation of the SU2 Fluid Dynamics Solver for Isentropic Non-Ideal Compressible Flows. Bachelor thesis, Delft University of Technology, Leganés, Spain (2022)
10. Span, R.: Multiparameter Equations of State. Springer, Berlin Heidelberg, Bochum, Germany (2000)
11. Thol, M., Dubberke, F.H., Rutkai, G., Windmann, T., Köster, A., Span, R., Vrabec, J.: Fundamental equation of state correlation for hexamethyldisiloxane based on experimental and molecular simulation data. Fluid Phase Equ. **418**, 133–151 (2016)
12. Economon, T.D., Palacios, F., Copeland, S.R., Lukaczyk, T.W., Alonso, J.J.: SU2: an open-source suite for multiphysics simulation and design. AIAA J. **54**(3), 828–846 (2016)
13. Pini, M., Vitale, S., Colonna, P., Gori, G., Guardone, A., Economon, T., Alonso, J.J., Palacios, F.: SU2: the open-source software for non-ideal compressible flows. J. Phys.: Conf. Ser. **821**, 012013 (2017)
14. Gori, G., Zocca, M., Cammi, G., Spinelli, A., Guardone, A.: Experimental assessment of the open-source SU2 CFD suite for ORC applications. Energy Proc. **129**, 256–263 (2017)
15. Head, A.J., Iyer, S., De Servi, C., Pini, M.: Towards the validation of a CFD solver for non-ideal compressible flows. Energy Proc. **129**, 240–247 (2017)
16. Bills, L.: Validation of the SU2 Flow Solver for Classical Non Ideal Compressible Fluid Dynamics. Master's thesis, Delft University of Technology, Delft, The Netherlands (2020)
17. Blanca Fuentes Monjas, Adam J. Head, Carlo De Servi, and Matteo Pini. Validation of the SU2 Fluid Dynamic Solver for Isentropic Non-Ideal Compressible Flows. In Martin White, editor, *Proceedings of the 4th International Seminar on Non-Ideal Compressible Fluid Dynamics for Propulsion and Power*, volume 29, pages 91–99, London, United Kingdom, 2022
18. Spinelli, A., Pini, Matteo, D., Vincenzo, G., Paolo, C., Francesco: design, simulation, and construction of a test rig for organic vapors. J. Eng. Gas Turbines and Power **135**(4), 042304 (2013)
19. Head, A.J., Michelis, T., Beltrame, F., Fuentes-Monjas, B., Casati, E., De Servi, C., Colonna, P.: Mach number estimation and pressure profile measurements of expanding dense organic vapors. In: White, M. (ed.) Proceedings of the 4th International Seminar on Non-Ideal Compressible Fluid Dynamics for Propulsion and Power, vol. 29, pp. 229–238, London, United Kingdom (2022)
20. Bell, I.H., Wronski, J., Quoilin, S., Lemort, V.: Pure and pseudo-pure fluid thermophysical property evaluation and the open-source thermophysical property library coolprop. Ind. Eng. Chem. Res. **53**(6), 2498–2508 (2014)
21. Weith, T., Heberle, F., Preissinger, M., Bruggemann, D.: Performance of siloxane mixtures in a high-temperature organic rankine cycle considering the heat transfer charactersitics during evaporation. Energies **7**, 5548–5565 (2014)

22. Preissinger, M., Bruggemann, D.: Thermal stability of hexamethyldisiloxane (MM) for high-temperature organic rankine cycle (ORC). Energies **9**(3), 183 (2016)
23. Michelis, T., Head, A.J., Majer, M., Colonna, P., De Servi, C.: Assessment of particle image velocimetry applied to high-speed organic vapor flows. Exp. Fluids **65**(6), 90 (2024)
24. Thol, M.: Empirical multiparameter equations of state based on molecular simulation and hybrid data sets. Ph.D. thesis, Ruhr-Universität Bochum, Bochum, Germany (2016)

Hybrid Lattice Boltzmann Method for Non-Ideal Gases: Application to the Compressible Taylor-Green Vortex Flow

Lucien Vienne, Alexis Giauque, and Emmanuel Lévêque

Abstract Focusing on the Taylor-Green vortex flow, this study demonstrates the capability of the lattice Boltzmann method to simulate compressible turbulent dynamics of non-ideal dense gases. The numerical scheme relies on a hybrid approach coupling the lattice Boltzmann algorithm with a conservative finite-volume discretization for the total energy. In particular, simulations at an initial Mach number of one reveal notable differences in the temporal evolution of the dilatation field within supersonic regions, as compared to an equivalent perfect gas simulation.

Keywords Dense gases · Lattice Boltzmann method · Compressible turbulence

1 Introduction

In compressible flows, the fluid dynamics are highly dependent on the thermodynamic state. In the vicinity of the critical point, gases with large molecular complexity deviate significantly from the ideal gas assumption. For such dense gases, the flow is governed in part by the fundamental derivative of gas dynamics [1]

$$\Gamma = 1 + \frac{\rho}{c}\frac{\partial c}{\partial \rho}\bigg|_s, \tag{1}$$

where ρ is the fluid density, $c = \sqrt{\partial p/\partial \rho|_s}$ the speed of sound, p the pressure and s the entropy. Non-classical compressibility effects, such as expansion shock waves, have been numerically simulated when $\Gamma < 0$. Peculiar behaviors of compressible

L. Vienne (✉) · A. Giauque · E. Lévêque
CNRS, Ecole Centrale de Lyon, INSA Lyon, Universite Claude Bernard Lyon 1, Écully, France
e-mail: lucien.vienne@ec-lyon.fr

A. Giauque
e-mail: alexis.giauque@ec-lyon.fr

E. Lévêque
e-mail: emmanuel.leveque@ec-lyon.fr

A. Giauque et al. (eds.), *Proceedings of the 5th International Seminar on Non-Ideal Compressible Fluid Dynamics for Propulsion and Power*, ERCOFTAC Series 32,
https://doi.org/10.1007/978-3-031-92695-2_7

non-ideal gas dynamics have attracted attention [2, 3], especially in energy conversion applications where heat transfer fluids may have this high molecular complexity. High fidelity simulations, such as large eddy simulations, offer valuable insights into real gas turbomachinery flows but are computationally expensive [4]. The lattice Boltzmann (LB) method is an alternative tool for solving the Navier-Stokes equations. Thanks to its algorithmic simplicity, the LB method has the potential to significantly reduce the computational cost of simulating flows in complex industrial configurations. Originally limited to weakly compressible flows, recent progress have extended the application range of the LB method to include compressible regimes. Similarly to the work of Reyhanian et al. [5], this article extends the LB method to compressible non-ideal flow dynamics.

We intend to achieve two objectives. First, we aim to adapt the hybrid compressible lattice Boltzmann of [6], which couples the LB method for the fluid density and momentum with a finite-volume scheme for the energy, to handle non-ideal gases. The finite-volume scheme proposed by [7] is employed to ensure a truly conservative scheme for the total energy when combined with the LB method. Secondly, we showcase the compressible Taylor-Green vortex flow and highlight differences when using a dense gas with $\Gamma < 0$.

2 Methodology

2.1 LB Scheme for Mass and Momentum

The LB scheme governs the time evolution of the distribution functions $(f_0, \ldots, f_{N-1})$ of particles with kinetic velocities $(\boldsymbol{e}_0, \ldots, \boldsymbol{e}_{N-1})$. At the macroscopic level, flow variables are recovered by summing the contributions from the distribution functions and calculating the moments so that

$$\rho = \sum_{i=0}^{N-1} f_i \quad \text{and} \quad \rho u_\alpha = \sum_{i=0}^{N-1} f_i e_{i\alpha}, \tag{2}$$

where ρ and u_α represent respectively the mass density and the velocity of the fluid. Reducing the kinetic velocity space to a finite basis of Hermite polynomials, and resorting Gaussian quadrature formula [8, 9], transforms the Boltzmann equation to the discrete-velocity Boltzmann equation. In this study, the velocity space is truncated to the standard D3Q27 lattice set with $N = 27$ distribution functions in three dimensions. This stencil encompasses all the adjacent nodes in the uniform Cartesian grid as shown in Fig. 1.

The discretization of the physical space and time involves the integration along the characteristics of the discrete-velocity Boltzmann equation, and results in the lattice Boltzmann scheme, a two-step algorithm that consists of

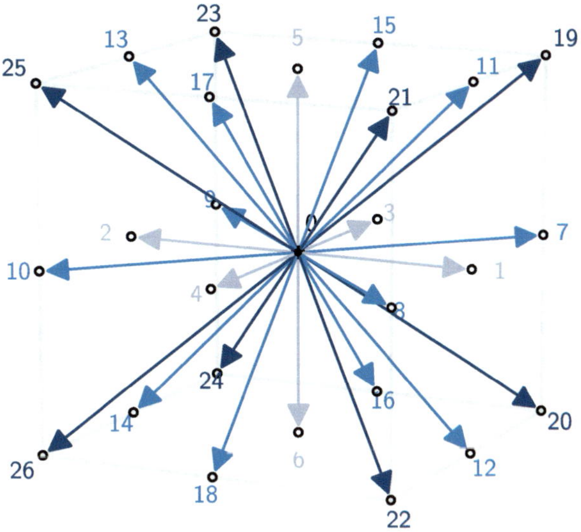

Fig. 1 Sketch of the D3Q27 stencil. The current node is the black-filled circle. Neighboring nodes are drawn as black non-filled circles. D3DQ27 distribution functions correspond to all possible permutation of $e_{i\alpha} = 0, -1, +1$ for $i = 0, \ldots, 26$ and $\alpha = x, y, z$

$$f_i^{coll}(t, \boldsymbol{x}) = f_i^{eq}(t, \boldsymbol{x}) + \left(1 - \frac{\Delta t}{\tau}\right) f_i^{neq}(t, \boldsymbol{x}) + \frac{\Delta t}{2} F_i(t, \boldsymbol{x}) \qquad \text{collision step,} \tag{3}$$

$$f_i(t + \Delta t, \boldsymbol{x}) = f_i^{coll}(t, \boldsymbol{x} - \boldsymbol{e}_i \Delta t) \qquad \text{streaming step,} \tag{4}$$

for $i = 0, \ldots, N$. Usual notations for the LB scheme are employed. τ is the relaxation time toward the equilibrium distribution function f_i^{eq} and is related to the viscosity. F_i refers to the source term, and the non-equilibrium part reads

$$f_i^{neq} = f_i - f_i^{eq} + \frac{\Delta t}{2} F_i \tag{5}$$

The equilibrium distribution stems from the Maxwell-Boltzmann distribution, and its discrete formulation is a cubic polynomial involving only local values of the fluid density and velocity. To improve the stability, the non-equilibrium part of the distribution functions are filtered following the recursive regularized collision procedure only [10], corresponding to $\sigma = 1$ in [6]. We also employ an artificial bulk viscosity based on the Mach number as proposed in [6].

As shown in [6], the aim of the source term F_i is to introduce a correction term $Corr_{\alpha\beta}$ that is essential for deriving the right equivalent macroscopic equations. Using the Chapman-Enskog expansion, one gets

$$\begin{aligned} Corr_{\alpha\beta} &= \delta_{\alpha\beta}\partial_t\left[p(\rho, T) - \rho c_s^2\right] - u_\alpha \partial_\beta \left[p(\rho, T) - \rho c_s^2\right] - u_\beta \partial_\alpha \left[p(\rho, T) - \rho c_s^2\right] \\ &\quad - \partial_\gamma D_{\alpha\beta\gamma}^{eq} + \delta_{\alpha\beta}\frac{2}{3}\rho c_s^2 \partial_\gamma u_\gamma . \end{aligned} \tag{6}$$

This expression corresponds to Eq. (52) and Eq. (54) in reference [6] under the assumption of an ideal gas law $p = \rho\theta c_s^2$ and with $\kappa = 0$. In Eq. (6), the corrections terms are discretized using finite difference schemes, employing a first-order backward scheme for the temporal derivative, a second-order central scheme for the divergence operator, and a first-order upwind scheme for the remaining spatial derivatives.

Except for leaving the pressure equation undefined and using the D3Q27 velocity set, the present LB scheme is similar to the one proposed by Farag et al. in [6]. This previous study opted for the entropy formulation for energy discretization due to its inherent property of decoupled linear stability. We adopt here the strategy proposed more recently by Wissocq et al. [7] for a total energy formulation. This conservative form of the energy scheme ensures the recovery of the correct shock speed.

2.2 Finite-Volume Scheme for Total Energy

Adapting Riemann solvers to the LB formalism, Wissocq et al. [7] introduced a conservative scheme that is linearly equivalent to its non-conservative entropy formulation. In particular, this scheme preserves the low dissipation characteristic of the LB method for isentropic phenomena such as acoustic propagation and vorticity. The conservative formulation for the total energy is expressed as

$$\partial_t (\rho E) + \partial_\alpha \mathcal{F}_\alpha^{\rho E} = 0, \tag{7}$$

where $\mathcal{F}_\alpha^{\rho E}$ denotes the total energy flux, encompassing both advective and diffusive components. Detailed expressions of these fluxes are provided in the study by Wissocq et al. [7]. These expressions depend not only on the macroscopic quantities (ρ, u_α, E) and the flux discretization scheme used, but also on the distribution functions f_i. Additional terms in $\mathcal{F}_\alpha^{\rho E}$ ensure consistency between the mass and momentum (solved by LB method) and the total energy solved by finite-volume method. The same computational grid as the LB method is used. The MUSCL-Hancock method MHM2D is selected as suggested in [7].

2.3 Thermal and Caloric Equation of States/Transport Coefficients

The governing equations require a model to establish the relationship between the thermodynamic variables p, ρ, and e. We need to provide the thermal and caloric equations of states, respectively:

$$p = p(\rho, T) \quad \text{and } e = (\rho, T)\,. \tag{8}$$

T denotes temperature, and e represents internal energy, which is related to total energy through $E = e + \frac{1}{2}u_\alpha u_\alpha$. Both thermal and caloric equations of state are linked via the relation

$$e = e_{ref} + \int_{T_{ref}}^{T} c_{v_\infty}(T')\,dT' + \int_{\rho_{ref}}^{\rho} \left[T\left(\frac{\partial p}{\partial T}\right)_\rho - p \right] \frac{d\rho'}{\rho'^2}, \tag{9}$$

where the $_{ref}$ subscript refers to an arbitrary reference state, c_{v_∞} is the specific heat at constant volume in the ideal gas limit. The last term in Eq. (9) represents the deviation from the assumption of a calorically perfect gas and is contingent upon the equation of state. Following [11], the ideal isochoric heat capacity is approximated by a power law of the form

$$c_{v_\infty}(T) = c_{v_\infty}(T_c)\left(\frac{T}{T_c}\right)^n, \tag{10}$$

with T_c denoting the critical temperature, and $c_{v_\infty}(T_c)$ and n being dependent on the simulated gas.

In this study, we consider two thermal equations. The simplest relation, known as the ideal gas assumption, is valid in the dilute gas region:

$$p(\rho, T) = \rho R_g T, \tag{11}$$

where R_g denotes the specific gas constant. More complex yet more accurate equations of states are necessary to explore non-classical gas dynamic phenomena realistically. The Peng-Robinson equation of state is valid in the thermodynamic region where non-classical gas phenomena are expected to be observed for heavier substances [12]. It is expressed as

$$p(\rho, T) = \frac{\rho R_g T}{1 - b\rho} - \frac{a(T)\rho^2}{1 + 2b\rho - b^2\rho^2}, \tag{12}$$

where

$$a = 0.45724\frac{R_g^2 T_c^2}{p_c}\alpha(T) \quad \text{and } b = 0.07780\frac{RT_c}{p_c}, \quad \text{with } \alpha(T) = \left(1 + \kappa(\omega)(1 - \sqrt{T/T_c})\right)^2. \tag{13}$$

The κ coefficient varies according to the accentric factor ω of the selected gas.

The gas viscosity and thermal conductivity must be described according to the thermodynamic variables. When the perfect gas assumption is involved, the Sutherland law for air and a constant Prandtl number equal to 0.71 are used. For dense gas simulations, we employ the generalized laws for viscosity and thermal conductivity of Chung et al. [13].

2.4 Taylor Green Vortex Flow Configuration

Due to nonlinear interactions, the Taylor-Green vortex (TGV) flow evolves from laminar to turbulent states, thereby illustrating in the process the Kolmogorov's cascade phenomenon. It provides insights into turbulent flows and serves as a benchmark for testing numerical schemes. Traditionally incompressible or at low Mach number, recent studies have extended the TGV configuration to compressible flows, adding complexity with shock waves. Additional details and validations are provided in [14]. In this study, we consider the TGV flow configuration in the context of a non-ideal gas in the inversion region $\Gamma < 0$ at an initial Mach number of one.

The initial condition consists of counter-rotating vortices in a three-dimensional periodic box of size $0 \leq x, y, z \leq 2\pi L$. It expresses as

$$u_x(x, y, z, t = 0) = U_0 \sin(\frac{x}{L}) \cos(\frac{y}{L}) \cos(\frac{z}{L}), \tag{14}$$

$$u_y(x, y, z, t = 0) = -U_0 \cos(\frac{x}{L}) \sin(\frac{y}{L}) \cos(\frac{z}{L}), \tag{15}$$

$$u_z(x, y, z, t = 0) = 0. \tag{16}$$

The simulation is initialized with constant pressure p_0 and temperature T_0. For the perfect gas simulation, the operating point is set at $\rho_0 = 1.204$ [kg/m^3] and $T_0 = 293.15$ [K]. To highlight potential non-ideal effects, the operating point of the dense gas simulation is chosen within the inversion region ($\Gamma < 0$) with $p_0 = 0.96 p_c$ and $\rho_0 = \rho_c/1.65$, resulting in a compressibility factor $Z_0 = p_0/(\rho_0 R_g T_0) \approx 0.43$. Thermodynamic properties of the ideal gas (air) and non-ideal gas (FC70, $C_{15}F_{33}N$) are presented in Table 1. The initial Mach number Ma $= 1$ and Reynolds number Re $= 1600 = \rho_0 U_0 L/\mu_0$ set the value of U_0 and L respectively. A grid sensitivity study yielded an optimal resolution of 768^3 grid points, with an initial Courant-Friedrichs-Lewy number of 0.49.

Table 1 Thermodynamic properties of the fluids

R_g [J.kg^{-1}K^{-1}]	γ [−]
287.05	1.4

(a) *Perfect gas (air)*

$\mathcal{M}$ [kg.mol^{-1}]	T_c [K]	p_c [atm]	$c_{v_\infty}(T_c)/R_g$ [−]	Z_c [−]	n [−]	ω [−]
0.821	608.2	10.2	118.7	0.270	0.4930	0.7584

(b) *Dense gas (FC70)*

3 Results

3.1 Compressible Taylor-Green Vortex Flow in a Non-Ideal Gas

The TGV simulations are advanced in time until twenty convective times $t^* = tU_0/L$. Figure 2 shows the temporal evolution of the dimensionless kinetic energy, the dimensionless enstrophy (solenoidal viscous dissipation) integrated over the entire domain, as well as that of the maximum Mach number within the domain. The initial decrease in total kinetic energy during the first convective (or identically acoustic) time suggests a transfer from kinetic to internal energy due to the out-of-equilibrium initial constant pressure field. Conversely, the kinetic energy increases during the early stage $2 < t^* < 4$ of the compressible TGV flow, consistently with previous observations [15, 16]. Two small peaks appear later indicating subsequent conversions from internal to kinetic energy. The evolution of the kinetic energy and enstrophy for the ideal and non-ideal gas simulations is similar, except for the presence of an enstrophy plateau in the ideal gas case compared to a peak in the non-ideal case. In contrast to the two previous plots, the evolution of the maximum Mach number differs significantly until $t^* \approx 9$ between the ideal and non-ideal gas simulations.

In Fig. 3, we plot the Mach number at time $t^* = 9.5$ on a slice located at $x = L/5$. Four main symmetric structures are visible, appearing very similar in both ideal and non-ideal gas simulations. The supersonic flow occurs only in small localized regions. Figure 4 shows the dilation field $\theta = \partial_\alpha u_\alpha$ normalized by its standard deviation at time $t^* = 9.5$ on a slice $x = L/5$. The color-map is saturated at ± 5 to highlight strong compression and expansion events. Contrary to the Mach field, the dilation field is distinct between the ideal and the non-ideal gas. In particular, thin sheets of moderate compression are noticeable in the ideal gas. As observed in homogeneous isotropic turbulence [17, 18], large expansions (red spots where $\theta/\theta_{STD} > 5$) are only present in the non-ideal gas simulation, inside the supersonic regions.

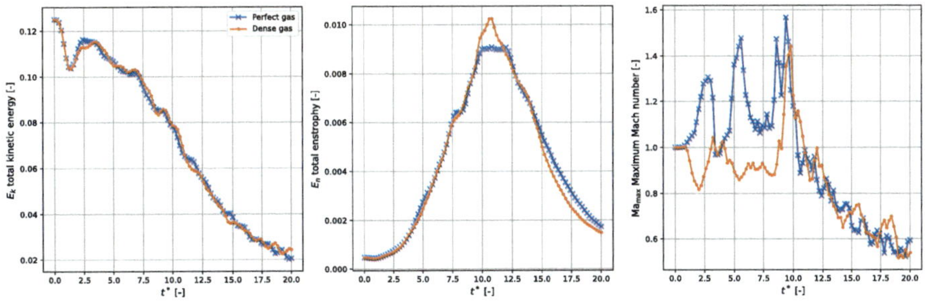

Fig. 2 Total kinetic energy, total enstrophy, and maximum Mach number for perfect gas and dense gas compressible Taylor-Green vortex

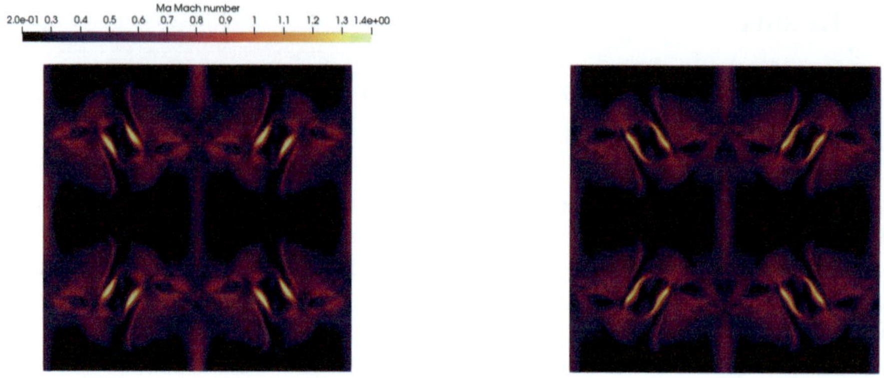

Fig. 3 Mach number distribution at time $t^* = 9.5$ on a slice located at $x = L/5$. Left: ideal gas. Right: non-ideal gas

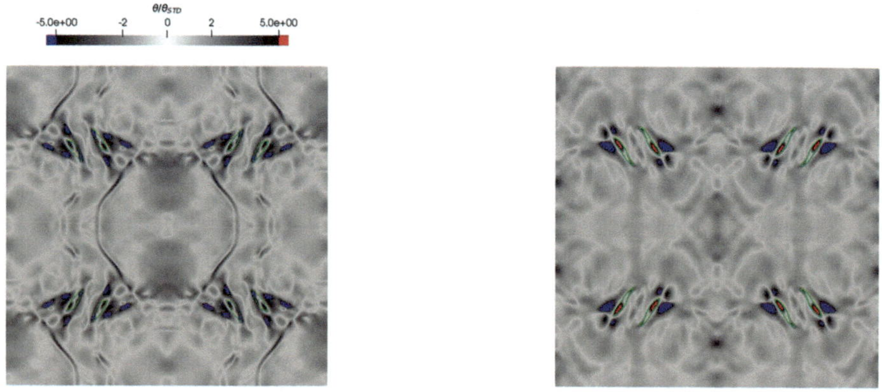

Fig. 4 Normalized dilation field at time $t^* = 9.5$ on a slice located at $x = L/5$. The green curve represents the isocontour of $\mathrm{Ma} = 1$. Left: ideal gas. Right: non-ideal gas

4 Conclusions and Future Works

This study uses Farag et al.'s hybrid compressible lattice Boltzmann method for non-ideal gases [6] and Wissocq et al.'s finite volume scheme for total energy [7], ensuring a truly conservative energy formulation when coupling these methods. The simulations focus on the well-known Taylor-Green vortex flow, but in the compressible regime with an initial Mach number of one. We highlight some similarities and differences between the ideal and non-ideal simulations. Future works will concentrate on achieving stable simulations at higher Mach numbers. Increased compressibility effects should indeed reveal more differences between ideal and non-ideal gases.

Acknowledgements This work was granted access to the HPC resources of PMCS2I at École Centrale de Lyon, Écully, France. This project was provided with computer and storage resources by GENCI at CINES thanks to the grant 2023-AD012A14929 on the supercomputer Adastra equipped with the MI250x GPU cards.

References

1. Philip, A.T.: A fundamental derivative in gasdynamics. Phys. Fluids **14**(9), 1843–1849 (1971)
2. Alberto, G., Piero, C., Matteo, P., Andrea, S.: Nonideal compressible fluid dynamics of dense vapors and supercritical fluids. Ann. Rev. Fluid Mech. **56**(1), 241–269 (2024)
3. Cinnella, P., Gloerfelt, X.: Insights into the turbulent flow of dense gases through high-fidelity simulations. Comput. Fluids **267**, 106067 (2023)
4. Alexis, G., Dominik, S., Christophe, C.: High-fidelity numerical investigation of a real gas annular cascade with experimental validation. Phys. Fluids **35**(12) (2023)
5. Reyhanian, E., Dorschner, B., Karlin, I.V.: Thermokinetic lattice boltzmann model of nonideal fluids. Phys. Rev. E. **102**(2) (2020)
6. Farag, G., Coratger, T., Wissocq, G., Zhao, S., Boivin, P., Sagaut, P.: A unified hybrid lattice-boltzmann method for compressible flows: Bridging between pressure-based and density-based methods. Phys. Fluids **33**(8) (2021)
7. Wissocq, G., Coratger, T., Farag, G., Zhao, S., Boivin, P., Sagaut, P.: Restoring the conservativity of characteristic-based segregated models: Application to the hybrid lattice boltzmann method. Phys. Fluids **34**(4) (2022)
8. Xiaowen, S., Xiaoyi, H.: Discretization of the velocity space in the solution of the boltzmann equation. Phys. Rev. Lett. **80**(1), 65–68 (1998)
9. Shan, X., Yuan, X.-F., Chen, H.: Kinetic theory representation of hydrodynamics: a way beyond the navierstokes equation. J. Fluid Mech. **550**(1), 413 (2006)
10. Orestis, M.: Increasing stability and accuracy of the lattice boltzmann scheme: Recursivity and regularization (2015)
11. Guardone, A., Argrow, B.M.: Nonclassical gasdynamic region of selected fluorocarbons. Phys. Fluids **17**(11) (2005)
12. Guardone, A., Vigevano, L., Argrow, B.M.: Assessment of thermodynamic models for dense gas dynamics. Phys. Fluids **16**(11), 3878–3887 (2004)
13. Ting, H.C., Mohammad, A., Lloyd, L.L., Kenneth, E.S.: Generalized multiparameter correlation for nonpolar and polar fluid transport properties. Ind. Eng. Chem. Res. **27**(4), 671–679 (1988)
14. Lucien, V., Alexis, G., Emmanuel, L.: Hybrid lattice boltzmann method for turbulent non-ideal compressible fluid dynamics. Phys. Fluid (2024). Under-review
15. Naifu, P., Yue, Y.: Effects of the mach number on the evolution of vortex-surface fields in compressible taylor-green flows. Phys. Rev. Fluids **3**(1) (2018)
16. David, J.L., Neil, D.S.: Assessment of low-dissipative shock-capturing schemes for the compressible taylorgreen vortex. AIAA J. **59**(2), 533–545 (2021)
17. Sciacovelli, L., Cinnella, P., Content, C., Grasso, F.: Dense gas effects in inviscid homogeneous isotropic turbulence. J. Fluid Mech. **800**, 140–179 (2016)
18. Giauque, A., Corre, C., Vadrot, A.: Direct numerical simulations of forced homogeneous isotropic turbulence in a dense gas. J. Turbul. **21**(3), 186–208 (2020)

Direct Simulations of Non-Ideal Supersonic Flows in the Trailing Edge Region

Marco Oliveti, Andrew P.S. Wheeler, and Giacomo Persico

Abstract Trailing edge losses constitute a substantial portion of the overall losses in transonic and supersonic turbines. These losses are significantly affected by the so-called base pressure, the pressure downstream of the blade trailing edge. The importance of accurately predicting this quantity in the turbine design has driven numerous experimental investigations aimed at developing appropriate correlations, mostly valid for ideal gases. However, technological limitations have hindered similar analyses for organic fluids. Aiming to provide a foundational basis for addressing this issue, this paper presents the preliminary phase of a numerical investigation of the base pressure problem in supersonic non-ideal flows. The study utilizes the DNS code *3DNS*, focusing on conditions representative of organic Rankine cycle applications, choosing the siloxane MM as the working fluid. This paper details the setup of the numerical investigation, including the definition of the numerical domain and working conditions. A preliminary RANS analysis is conducted to examine the effects of the Reynolds number on the generated flow field and to provide an initial solution for initializing the DNS simulations. Finally, the first results from DNS are reported, highlighting the main findings and outlining the next steps for this investigation.

Keywords Organic rankine cycles · Turbines performances · Base pressure · High fidelity simulations · DNS

M. Oliveti (✉) · G. Persico
Politecnico di Milano, Department of Energy, Milan, Italy
e-mail: marco.oliveti@polimi.it

G. Persico
e-mail: giacomo.persico@polimi.it

A. P.S. Wheeler
University of Cambridge, The Whittle Laboratory,
Cambridge, United Kingdom
e-mail: aw329@cam.ac.uk

A. Giauque et al. (eds.), *Proceedings of the 5th International Seminar on Non-Ideal Compressible Fluid Dynamics for Propulsion and Power*, ERCOFTAC Series 32,
https://doi.org/10.1007/978-3-031-92695-2_8

1 Introduction

The efficiency of Organic Rankine Cycles (ORCs) is heavily dependent on the performance of the expander [1], which, for medium to high power applications, is generally an axial or radial turbine [2]. Under specific operating conditions, these turbines, often designed to process high pressure ratios in a limited number of stages, generate transonic and supersonic flows characterized by complex shock patterns that significantly impact overall efficiency. One critical factor influencing the aerodynamic performance of these cascades is the so-called base pressure, the minimum pressure found in the region at the rear of a blade's trailing edge. Predicting this quantity is extremely complicated, as the flow near the trailing edge is heavily affected by turbulent shear stresses. As previously illustrated in [3], the base pressure directly influences trailing edge losses, which, in turn, represent the largest contribution to overall cascade losses when the flow is in transonic conditions. Understanding and optimizing base pressure are, therefore, crucial steps toward maximizing turbine efficiency. The relevance of the base region has already been assessed with air as a working fluid in a crucial experimental analysis carried out by Sieverding [4]. That study provided significant insights into the characterization of the base region and the flow around the trailing edge under different operating conditions. Experiments were performed with both a flat plate model, representative of the overhang section of the blade after the throat, and actual cascades, in an attempt to provide general correlations for base pressure prediction. Replicating such experiments with non-ideal flows poses additional challenges. Only a few facilities in the world [5–7], generally blowdown wind tunnels or Ludwig tubes, are designed to operate with organic fluids. However, even in these facilities, investigating the base region would prove to be extremely challenging. To achieve the required mass flow rates in lab-scale facilities, the geometry often needs to be miniaturised, complicating the analysis of the trailing edge area. Moreover, a proper evaluation of the flow field is aggravated by the need to design and implement measurement techniques that can operate effectively with organic fluids [8]. Given these complexities, numerical investigations offer a more feasible alternative for studying the base pressure in organic flows. The complexity of the flow field around the base region in transonic flows, caused by the onset of shock waves and flow separations, makes traditional Reynolds-averaged Navier-Stokes simulations (RANS) often unable to accurately capture these phenomena, necessitating a more advanced computational approach. An extended investigation through Direct Numerical Simulations (DNS), which can be considered a virtual alternative to experiments, was therefore conceived, aiming to provide a more precise understanding of the base pressure dynamics in organic flows. The research is conducted using the software *3DNS*, developed by Wheeler [9]. Due to the complexity of the research, the present paper only describes the setup of the case, including the definition of the domain, a preliminary RANS analysis, the meshing process and the first DNS results.

2 Domain Definition

The geometry employed for this research is inspired by the test rig designed by Sieverding for the experimental campaign described in [10]. In this setup, a flat plate with a rounded trailing edge was placed at the centreline of an asymmetric nozzle. The curvature of the nozzle was designed so that the upper throat would be located at the trailing edge coordinate, while the lower throat could be arbitrarily moved upstream. This configuration generates different flow accelerations on the two sides of the plate, creating a high- and a low-pressure side. This layout aims to replicate the pressure and suction sides of a turbomachinery blade, where the pressure side is characterised by an outlet Mach number of 1, and the suction side can achieve higher outlet Mach numbers depending on the position of the lower throat. The velocities at the end of the two sides of the plate directly affect the resulting flow field downstream of the trailing edge, producing specific base region configurations depending on the flow conditions.

However, reproducing the entire test rig geometry with a mesh suitable for DNS analysis is computationally prohibitive. Therefore, a simplified domain was devised to produce a similar flow field, merging the plate and the converging nozzle into a single profile, as depicted in Fig. 1.

The layout designed for the analysis is shown in the centre of the sketch, with two additional channels added with dashed lines to better visualize the periodicity of the domain. As the nozzle is no longer present, flow expansion is achieved by defining a convergent pressure side, allowing the flow to reach sonic conditions just before the plate trailing edge. Conversely, the suction side is perfectly flat, permitting further acceleration after the throat and reaching an outlet Mach number greater than 1, thus generating an asymmetric pressure distribution across the trailing edge. The resulting flow field configuration is function of the stagger angle β: the wider the angle, the less the flow is accelerated on the suction side, up to the limit condition at $\beta = 90°$, where the outlet Mach number is 1 on both sides. The domain dimensions are consistent with those set by Sieverding, with a trailing edge thickness of 5 mm, a throat section of 50 mm, and a plate length of 300 mm. These dimensions achieve a ratio between the boundary layer momentum thickness at the end of the plate and the trailing edge thickness representative of actual cascade applications. This length

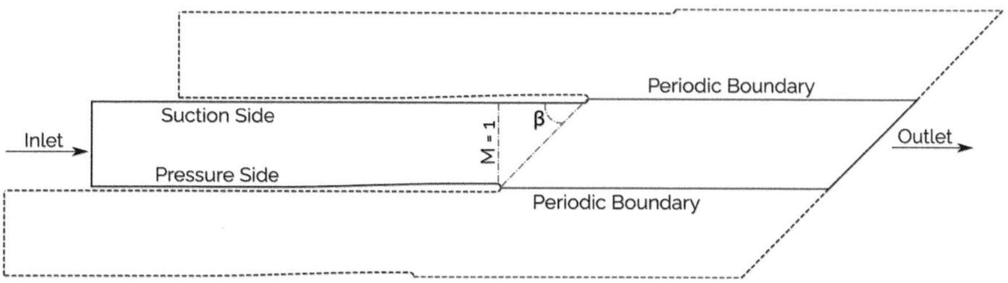

Fig. 1 Computational fluid domain

is also crucial for allowing a transition to a fully developed turbulent boundary layer from a stagnation point, as explained in Sect. 4. While a more extensive analysis could later be performed on the effect of the stagger angle on the resulting flow field, the current paper focuses on preliminary results obtained for a single configuration, where $\beta = 46°$. This specific condition was selected to achieve a high outlet Mach number on the plate suction side while avoiding interactions between fishtail shocks and the boundary layer, phenomena that could be analyzed in future works. Additionally, in [4], experimental data are reported for this specific layout regarding pressure development in the wake, which could be used for a numerical assessment of the results.

3 Boundary Conditions

As already specified, the present numerical investigation is carried out with both air and an organic fluid. Simulations with air are performed to validate the computational approach and the obtained flow field. These results can be compared with Sieverding's experimental data and the preliminary RANS simulations, whose results based on the ideal gas model are considered more reliable than those one obtained with the non-ideal working fluid. For the organic fluid, among the different possibilities commonly employed in ORCs, hexamethyldisiloxane (MM) was selected for this specific case study. This siloxane is often considered for medium to high temperature energy sources in machines where high expansion ratios are required, often generating transonic and supersonic flows, and where the dense gas behaviour is more noticeable than with other fluids.

Different inlet boundary conditions were chosen for the two fluids. For air, the same conditions used by Sieverding were maintained, with the inlet total temperature $T_{T0} = 280$ K and the inlet total pressure $p_{T0} = 2.3$ bar. It is worth mentioning, however, that the outlet Mach number and the resulting flow field would not change with different conditions, being exclusively functions of the area ratio between the inlet and the throat and of the specific heat ratio γ, which is constant for air. For the MM simulations, where the flow field is, instead, dependent on the upstream stagnation conditions, T_{T0} was set at 503 K and p_{T0} at 10 bar. This specific state was selected to mimic common operating conditions for this working fluid, while still investigating an expansion occurring in the non-ideal gas region, characterised by a compressibility factor of $Z_{T0} = 0.75$ and a fundamental derivative of thermodynamics $\Gamma = 0.74$. The outlet pressure, being the flow fully supersonic downstream of the throat, is determined by the upstream total quantities and by the area ratio: a sufficiently low initial pressure is imposed at the outlet as a starting value to induce the choking condition. After some timesteps, when the flow becomes completely supersonic downstream of the throat, this pressure is automatically adapted to the value that generates an ideal expansion, avoiding the generation of oblique shocks in the outlet region. The isentropic expansion for MM up to the actual outlet pressure is depicted in Fig. 2, along with the fluid saturation line and a contour of the

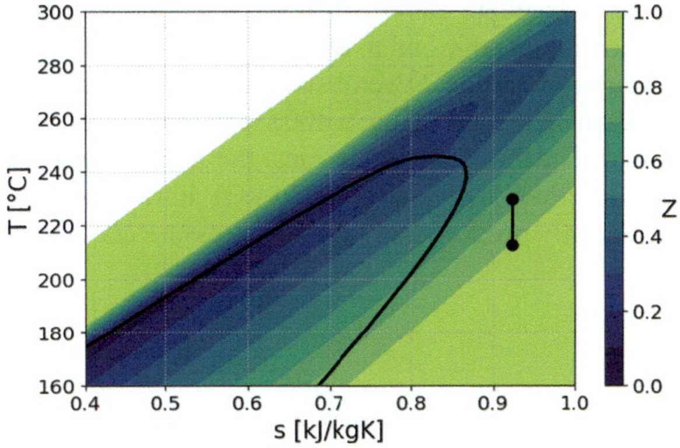

Fig. 2 MM expansion on TS diagram with Z contour

compressibility factor. Due to the non-ideality of the flow in the selected region, an appropriate thermodynamic model to generate equations of state for dense-vapour is required. To achieve thermodynamic consistency, the strategy implemented in the current version of *3DNS* is based on the exploitation of an entropy functional form which, coupled with fluid-specific coefficients extracted from the REFPROP library [11], provides the needed equation of states. A thorough description of the method can be found in [12].

4 Rans Preliminary Analysis

A critical parameter to define was the Reynolds number at which the simulations would be conducted. In this paper, the Reynolds number is defined as shown in Eq. 1,

$$Re = \frac{\rho_{is} v_{is} te}{\mu_{is}} \tag{1}$$

where the subscript *is* indicates quantities computed at the outlet through an isentropic expansion from the inlet stagnation quantities and the outlet static pressure. The characteristic length chosen for the Reynolds number, given the focus on the base region, is the trailing edge thickness *te*. The Reynolds number has a dual significance in this investigation. On one hand, the rapid increase in the number of mesh cells required for DNS simulations [13] makes it impossible to simulate organic flows under the described conditions, where the Reynolds number would reach approximately 1.5M, with the current computational resources. On the other hand, an excessively low Reynolds number would inhibit the laminar-to-turbulent transition, preventing the boundary layer from fully developing turbulence and resulting in a flow field that is not representative of actual turbomachinery operating conditions. Therefore, a compromise between these two competing factors was necessary. A preliminary RANS analysis was conducted on the same 2D computational domain with both fluids to evaluate the effect of the Reynolds number on the flow field downstream of the trailing edge and to determine the minimum value required to achieve boundary layer transition. Several simulations were performed using ANSYS Fluent, with the domain meshed using a 400 k-cell multiblock structured grid created in ANSYS ICEM. For the MM simulations, the real gas model was introduced using look-up tables defined through the REFPROP library. As shown in [3], trailing edge losses are directly proportional to a base pressure coefficient, defined as reported in Eq. 2.

$$C_{b,p} = \frac{p_b - p_{ref}}{\frac{1}{2}\rho_{ref} v_{ref}^2} \tag{2}$$

To evaluate the base pressure and the reference conditions, the two lines depicted in Fig. 3, which illustrates the resulting pressure contour from a simulation performed with air, were extracted from each simulation.

The reference conditions are taken from the channel centrepoint, on the line that connects the two trailing edges. As far as the base pressure is concerned, instead, it can be seen from Fig. 4 that the region immediately downstream of the trailing edge simulated by the RANS does not present a constant pressure. The value p_b was chosen as the minimum pressure from the represented curve, as per definition of base pressure.

The inclination of the p_b line was adjusted for every simulation according to the resulting wake angle. The $C_{b,p}$ coefficient was computed for both fluids in several simulations, each characterized by a different Reynolds number, which was obtained by scaling the Sutherland reference dynamic viscosity for air and the geometry for MM. The results are presented in Fig. 5.

As illustrated in the left plot of Fig. 5, the base pressure coefficient, $C_{b,p}$, shows significant variation at lower Reynolds numbers, with a very similar trend shared by the ratio between base pressure and inlet total pressure, in the right plot. However, as the Reynolds number increases, the variation of $C_{b,p}$ begins to stabilize, reaching a region where the coefficient changes linearly with Re and does not vary excessively

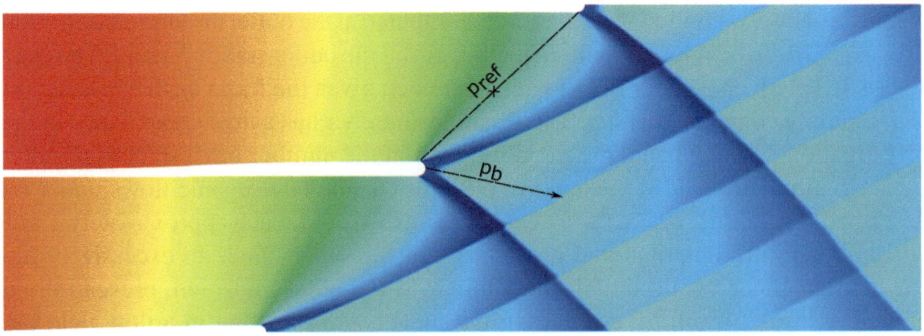

Fig. 3 Pressure contour from a RANS simulation in air. The two lines from where the reference and the base conditions are extracted have been included

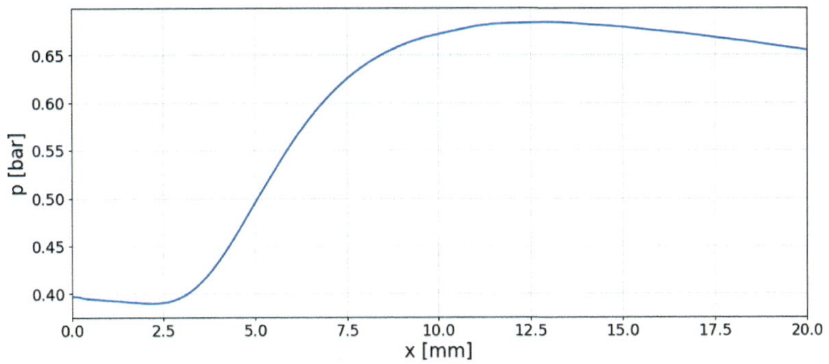

Fig. 4 Pressure trend extracted from a RANS simulation downstream of the trailing edge

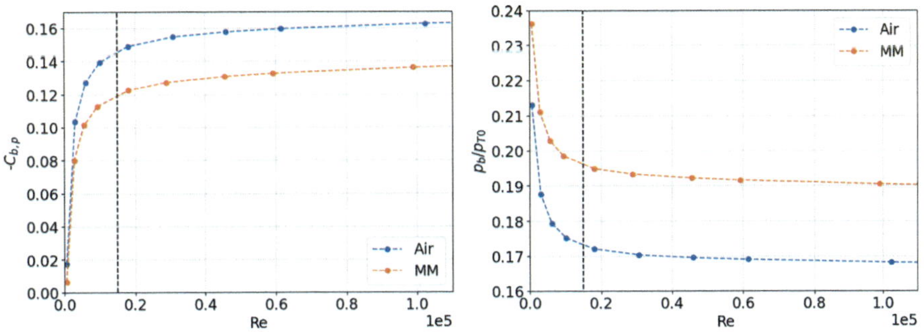

Fig. 5 Variation of base pressure coefficient and of base pressure with Reynolds number

at higher values. This behaviour indicates that beyond a certain Reynolds number, the base pressure coefficient becomes less sensitive to further increases in Re, which is essential for ensuring a realistic representation of the flow field downstream of the trailing edge. Re = 15k, highlighted in the plot through a vertical dashed line, was therefore selected as the appropriate value.

To ensure a fully developed turbulent boundary layer at the specified Reynolds number, a transition analysis using RANS simulations was conducted, aiming to verify that the flow transitions to a fully turbulent state for both fluids. The same setup with Ansys Fluent was employed, using the k-ω SST model coupled with the γ-Re_θ transition model [14] to simulate turbulence and detect whether and where the boundary layer would transition. From the simulations results, the displacement thickness δ^* and momentum thickness θ^* were computed by integrating the respective functions along the boundary layer thickness δ across the whole suction side. For the definition of δ, Eq. 3 was used, extracted from [15].

$$\delta = \frac{5}{2}\frac{\int_0^\infty y(|\frac{\partial v}{\partial y}| + |\frac{\partial T}{\partial y}|)dy}{\int_0^\infty (|\frac{\partial v}{\partial y}| + |\frac{\partial T}{\partial y}|)dy} \quad (3)$$

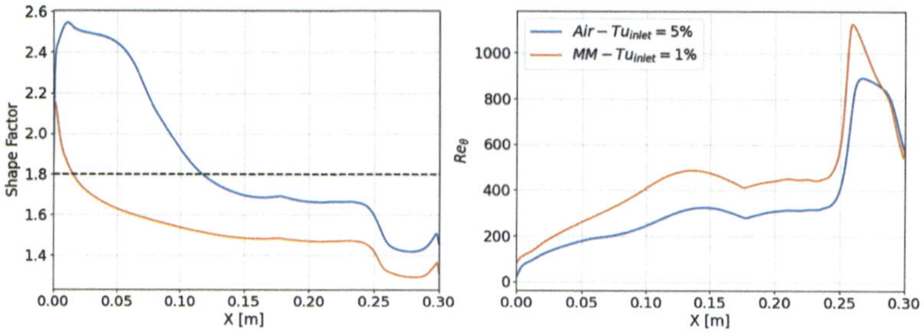

Fig. 6 Shape Factor and Re_θ as function of x coordinate on the suction side

The shape factor H was then computed to determine boundary layer transition, considering 1.8 as a threshold value below which the flow can be considered fully turbulent. The results for the two fluids are depicted in Fig. 6.

The two plots show the shape factor and the Reynolds number based on momentum thickness Re_θ as function of the x coordinate along the plate suction side. It can be seen that turbulent transition is reached early for MM, while air requires a larger distance to develop a fully turbulent boundary layer. The explanation for this phenomenon could be provided by the right plot, highlighting for the same value of Re, higher Re_θ for the organic fluid. It has to be noted also that, while for MM setting an inlet turbulence intensity of 1% was enough to reach the boundary layer turbulent development, the same quantity needed to be increased to 5% for air, which, even if slightly higher than typical values for wind tunnels, it is still acceptable for the current investigation.

5 Mesh

The definition of the mesh is one of the most crucial aspects of DNS setup. Explicit time stepping is used in this kind of compressible solvers, in order to avoid the additional computational burden of dual time stepping and implicit time integration. The employment of this numerical scheme requires, however, a fairly uniform grid, with very regular cells throughout the whole domain. This feature is achieved in *3DNS* by an iterative process that uses a Poisson solver to regularise the block edges, which are then equally split within the free flow and subdivided according to the necessary grading towards the walls. With this strategy, nearly squared cells are obtained in the free flow, while close to the walls a y^+ of 1 is reached, keeping x^+ and z^+ respectively below 15 and 10 in the finest mesh. The near-wall cell size is then particularly important due to the necessity to resolve the smallest scales of turbulence and also due to the definition of the Courant number, that determines the time step according to the minimum cell size. In the simulations later reported, a value of 0.8 was imposed as maximum Courant number to ensure stability. To

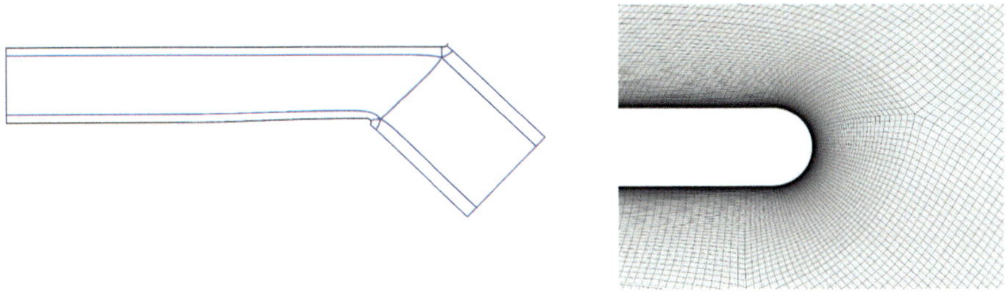

(a) *Blocking topology* (b) *Mesh detail - every 3rd line highlighted*

Fig. 7 Mesh definition

create an appropriate mesh for the domain described in Sect. 2, an apposite 8-blocks topology was conceived, as reported in Fig. 7a.

An o-grid structure is built around the plate, with a width fixed to 15 mm. The blocks are then split around the trailing edge to allow a better mesh refinement and good orthogonality. The outlet section is turned by the same angle β, in order to reduce the cells skewness in the region downstream of the plate. The upper and lower edges of this region are defined by periodic boundary conditions. In Fig. 7b, a detail of the grid obtained with around 2M cells in the plane is shown, illustrating the resulting mesh around the trailing edge when two periodic domains are plotted. This 2M cells mesh was used for the 2D reported cases, while the 3D simulations were carried out by extruding the mesh along a 0.03 m depth (representing 10% of the plate length) with 201 superimposed layers, reaching a total of around 400 M cells.

6 Results

In this last paragraph, some DNS results are presented. At first, the outcomes of four 2D simulations are reported and compared, while, later, two 3D investigations are described.

The 2D simulations were carried out setting a null inflow turbulence, in order to better visualize the effects of the Reynolds number and of the working fluid on the laminar flow, also avoiding the formation of intense structures, that would not be dissipated by the 3D turbulent breakdown. All the simulations, in order to ease the generation of a supersonic flow, were initialized from RANS results. The Mach contours of the trailing edge region extracted from four different simulations are reported in Fig. 8, considering both air and MM and two different Re, 15 k and 30 k. As for the RANS simulations, the scaling of the Reynolds number in 3DNS is achieved, for air, by tuning the Sutherland reference dynamic viscosity. For MM, a coefficient is instead introduced as a ratio between the actual viscosity computed

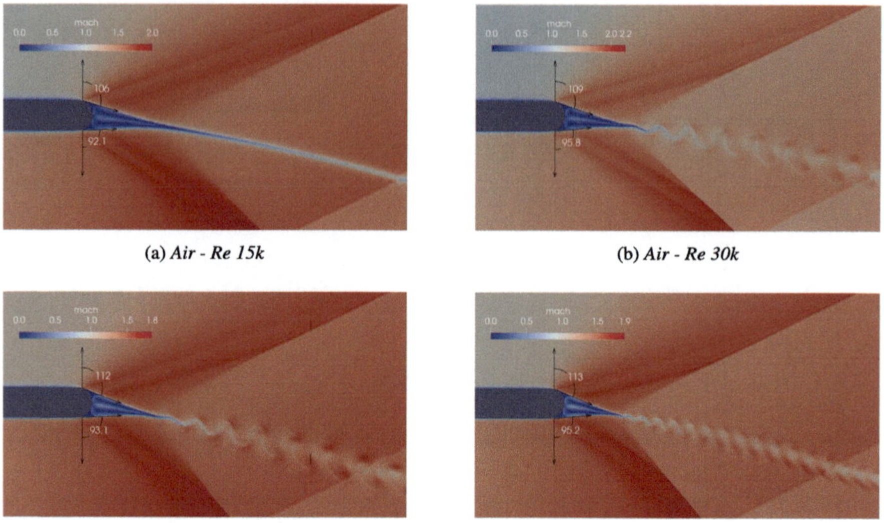

(a) *Air - Re 15k*

(b) *Air - Re 30k*

(c) *MM - Re 15k*

(d) *MM - Re 30k*

Fig. 8 2D simulations at different Reynolds numbers

by the non-ideal gas model and the simulated viscosity. To get a Re of 15 k, the coefficient was set as $\mu_{real}/\mu_{sim} = 0.01$.

Despite the preliminary nature of the presented simulations, some interesting insights can be derived from the figures. The contours reveal the formation of vortex structures in the wakes, even in the absence of inlet turbulence injection, in all cases except the lower Re air simulation. This phenomenon is dependent on both the fluid and the Reynolds number, indicating that wake instabilities are achieved at lower Re in MM than in air. As the Reynolds number increases, the frequency of these vortex structures also increases, with smaller vortex amplitudes and earlier dissipation observed at higher Re. The Mach number downstream of the trailing edge is consistently higher for air compared to MM. For air, this quantity is determined by the geometry and specific heat ratio, whereas in non-ideal flows like MM, the outlet Mach number is also influenced by the inlet boundary conditions. Both fluids exhibit a slight increase in exit Mach number with increasing Re. Although a higher Re leads to a smaller boundary layer width and consequently a smaller area ratio, the reduced base region at higher Re promotes a larger expansion in the area downstream of the trailing edge, resulting in a locally higher Mach number. This feature is coupled with the turning angle of the flow. Stronger shocks, and therefore greater Mach differences between the suction and pressure sides, are associated with a larger turning angle of the wake for the same fluid. Notably, MM exhibits a wider flow deflection even at lower Mach numbers compared to air. Additionally, the difference in flow deflection between the two Reynolds numbers is smaller for MM, indicating an asymptotic behaviour beyond a certain Re and demonstrating the validity of the selected conditions for achieving relevant results. Regarding the

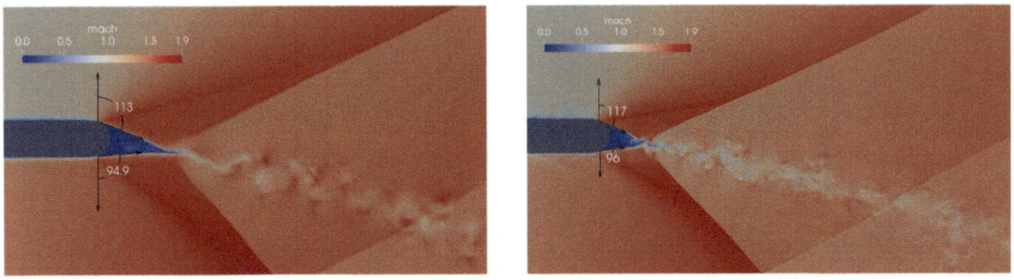

(a) *Null inlet turbulence intensity* (b) *2% inlet turbulence intensity*

Fig. 9 3D simulations results with MM–Re 15 k

base region, the traditional triangular shape is evident in the reported contours, with smaller regions observed for MM. The flow field features two counter-rotating recirculation bubbles, characterized by motion towards the outlet at the triangle edges and towards the trailing edge along the symmetry line. This structure, however, is a consequence of the bi-dimensional flow, as the lack of turbulent breakdown inhibits real mixing phenomena in the wake.

Moving to the 3D investigation, Fig. 9 reports two Mach contours extracted from MM simulations.

The only difference between them is the inlet turbulence intensity, with the simulation depicted in Fig. 9a still characterized, as the 2D cases, by a laminar inlet, and the one in Fig. 9b, featuring an inlet turbulence intensity of around 2%. Both simulations were carried out with Re = 15 k.

As evident from the contours, the wake flow in both cases is fully turbulent, lacking the regular vortex structures that have been disrupted by three-dimensional turbulence. However, a significant difference emerges between the two flow fields downstream of the trailing edge, where the base regions exhibit distinctly different shapes. In the first case, the plate boundary layer remains laminar until the trailing edge, causing the transition to turbulent flow to occur at the beginning of the wake itself. In contrast, in the second case, the boundary layer undergoes a turbulent transition shortly after the inlet, as accurately predicted by the RANS analysis presented in Sect. 4. This discrepancy between the two simulations is reflected in the longer base region observed in the case with zero turbulence intensity, resembling the 2D cases where no transition at all is occurring. Furthermore, when comparing the two wakes, it is evident that a stronger turbulent breakdown is present in the second case: although, in fact, the linear wake instability is already disrupted in the first simulation, a discernible oscillation frequency persists immediately downstream of the base region. In contrast, in the case with boundary layer transition, the entire wake is fully brokendown by turbulence and it is challenging to detect any periodicity. To properly visualize the boundary layer turbulent transition in the second simulation, the Q-Criterion was computed in the blocks representing the plate o-grids. Figure 10 reports its isosurface at a value of 10^8, superimposed to the Mach contour computed on a 2D slice at midspan. The picture well depicts the vortex struc-

Fig. 10 Q-Criterion on the MM simulation illustrating the boundary layer turbulent transition

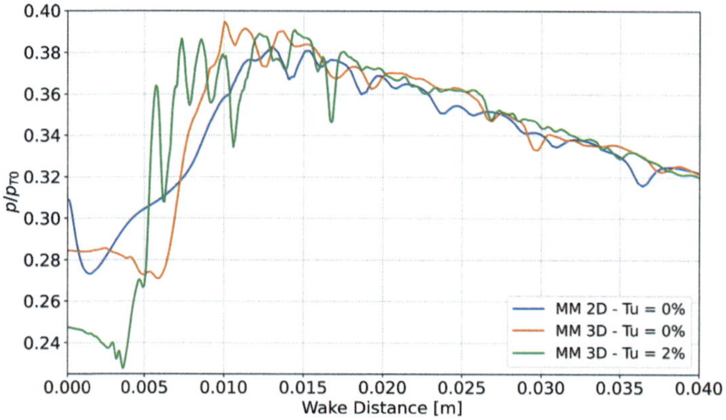

Fig. 11 Pressure trend along the wake in three MM simulations at Re = 15 k

tures originating near the wall a short distance after the inlet, developing a turbulent boundary layer much before the plate trailing edge.

Lastly, wake pressure was extracted from the three simulations with MM at Re = 15 k along the p_b line depicted in Fig. 3 and the comparison is shown in Fig. 11. As expected, the most significant differences are observed within the first 10 mm, where the base region is located, while very close values are found further downstream. The 2D simulation, represented in blue, lacks an isobaric region and exhibits a local pressure minimum shortly after the trailing edge, attributed to the specific flow configuration characterized by recirculation bubbles. In contrast, the 3D simulations, as also indicated by the Mach contours, display a more uniform base region where recirculation bubbles are absent, and the flow appears almost stationary throughout the triangular region. This is reflected in the plot by the presence of a short isobaric section, followed by a slight pressure decrease before the steep rise outside the base region. The plot also clearly illustrates the reduction in base region length when turbulent transition is introduced. This reduction is accompanied by a lower base pressure and a steeper pressure rise after reaching the minimum value. Additionally, the pressure oscillations provide valuable insights. In the 2D case, the onset of wake instabilities occurs approximately 10 mm downstream of the trailing edge, indicating an absence of flow disturbances before that point. The 3D simulation without inlet turbulence injection (shown in orange) exhibits the first oscillations in the middle of the base region, suggesting that turbulent transition occurs within the wake itself. Meanwhile, the 3D case with boundary layer transition (in green) shows oscillations starting immediately, confirming that the transition occurred upstream of the wake.

7 Conclusions and Future Works

This study has provided valuable insights into the high fidelity simulation of the base region in supersonic flows with non-ideal gases. Inspired by previous experimental campaigns conducted in air, a specific geometry was defined for DNS analysis, focusing on the base region. Appropriate boundary conditions were selected to analyze a first test case, and a preliminary RANS analysis was carried out to determine the Reynolds number effects on the flow. This helped in selecting a trade-off Reynolds number that allows the study of a case feasible with DNS and representative of actual operating conditions in turbomachinery. A high-quality mesh topology was developed, ensuring the accuracy required for DNS. The simulations included both 2D laminar DNS and 3D turbulent DNS with air and MM. These simulations highlighted the effects of Reynolds number and working fluid on the flow field downstream of the trailing edge, encompassing both the base region and the wake. The 3D DNS successfully achieved a fully turbulent wake and a complete turbulent transition in the plate boundary layer, marking a significant advancement beyond the initial 2D results. These efforts have laid a solid foundation for future investigations. The next steps will involve gathering detailed flow statistics to separate mean flow characteristics from turbulent effects. This will enhance the analysis of trailing edge losses and help identify the characteristic frequencies of vortex shedding. Additionally, similar simulations with air will be conducted to compare the results with Sieverding experimental data and to detect and understand the non-ideal gas effects. Further investigation could also explore different boundary conditions and stagger angles, with the ultimate aim of studying part-load conditions involving shock-boundary layer interactions.

Acknowledgements The author gratefully acknowledges the financial support from the KCORC (Knowledge Centre on Organic Rankine Cycle) for funding the secondment period at the Whittle Laboratory, University of Cambridge, where this work was developed. This support was instrumental in enabling the research and simulations presented in this study. We acknowledge the CINECA award under the ISCRA initiative, for the availability of high performance computing resources and support.

Appendix: Computational Time

The reported simulations were performed on the GPU-based European supercomputer Leonardo, hosted and managed by Cineca. The computational time for each simulation varied depending on the complexity of the mesh and the working fluid. The following table summarises the computational resources used, including mesh size, nodes, GPUs per node, and the time required for both 2D and 3D simulations. Data on 3D simulations in air have been included for comparison, although their results are not discussed in the current paper (Table 1).

Table 1 Summary of computational resources and time for DNS simulations on Leonardo HPC

Simulation type	Fluid	Mesh size [cells]	Nodes	GPUs per node	Time per step [s]
2D	Air	2M	1	1	$1.63 \cdot 10^{-2}$
2D	MM	2M	1	1	$1.76 \cdot 10^{-2}$
3D	Air	500M	2	4	0.75
3D	MM	500M	2	4	0.78

References

1. Ennio, M.: Theoretical basis of the organic Rankine cycle. In: Organic Rankine Cycle (ORC) Power Systems. Woodhead Publishing (2017)
2. Colonna, P., Casati, E., Trapp, C., Mathijssen, T., Larjola, J., Turunen-Saaresti, T., Uusitalo, A.: Organic rankine cycle power systems: From the concept to current technology, applications, and an outlook to the future. J. Eng. Gas Turb. Power **137**(10), 10 (2015)
3. David, B., John, J.O., Andrew, P.S.W.: The effect of Isentropic exponent on transonic turbine performance. J. Turbomach. **142**(8), 081007 (2020)
4. Sieverding, C., Stanislas, M., Snoeck, J.: The base pressure problem in Transonic turbine cascades. J. Eng. Power **102**(3), 711–718 (1980)
5. Andrea, S., Matteo, P., Vincenzo, D., Paolo, G., Francesco, C.: Design, simulation, and construction of a test rig for organic vapors. J. Eng. Gas Turb. Power **135**(4), 042304 (2013)
6. Mathijssen, T., Gallo, M., Casati, E., Nannan, N.R., Zamfirescu, C., Guardone, A., Colonna, P.: The flexible asymmetric shock tube (fast): a ludwieg tube facility for wave propagation measurements in high-temperature vapours of organic fluids. Exp. Fluids **56**(10), 195 (2015)
7. Francisco, J.D.G., Andrew, P.S.W., Jonathan, O.: A study of trailing-edge losses in organic rankine cycle turbines. J. Turbomach. **138**(12), 121003 (2016)
8. Camilla, C., Alberto, F., Andrea, S., Paolo, G., Alberto, G.: Pneumatic system for pressure probe measurements in transient flows of non-ideal vapors subject to line condensation. Measurement **192**, 110802 (2022)
9. Wheeler, A.P.S.: Desktop-dns: An open toolkit for turbomachinery aerodynamics,: Volume 13C: Turbomachinery–Deposition, Erosion, Fouling, and Icing. Ducts, Noise, and Component Interactions, Design Methods and CFD Modeling for Turbomachinery (2023)
10. Sieverding, C., Decuyper, M., Colpin, J., Amana, O.: Model tests for the detailed investigation of the trailing edge flow in convergent transonic turbine cascades (1976)

11. Lemmon, E.W., Bell, I.H., Huber, M.L., McLinden, M.O.: NIST standard reference database 23: Reference fluid thermodynamic and transport properties-REFPROP, version 10.0, National Institute of Standards and Technology (2018)
12. Wheeler, A.P.S.: High fidelity simulation of dense vapours with thermodynamic consistent modelling. Comput. Fluids **268**, 106088 (2024)
13. Sandberg, R.D., Michelassi, V.: The current state of high-fidelity simulations for main gas path turbomachinery components and their industrial impact. Flow Turbul. Combust. **102**(4), 797–848 (2019)
14. Langtry, R.B., Menter, F.R.: Correlation-based transition modeling for unstructured parallelized computational fluid dynamics codes. AIAA J. **47**(12), 2894–2906 (2009)
15. Giles, M.: UnsfloâŁ¯: A numerical method for the calculation of unsteady flow in turbomachinery. Massachusetts Institute of Technology. Gas Turbine Laboratory, Technical report (1991)

Multifidelity Optimization of an ORC Turbine Vane Using Scale-Resolving Simulations

Camille Matar, Paola Cinnella, and Xavier Gloerfelt

Abstract A multi-objective shape optimization of a high-pressure turbine (HPT) vane for Organic-Rankine Cycles (ORC) is designed by fusing together high-fidelity, scale-resolving simulations, and specifically wall-resolved Large Eddy Simulations (WRLES), and Reynolds-Averaged Navier Stokes (RANS) solutions through a multi-fidelity co-Kriging surrogate model. Given the extremely high cost of WRLES at Reynolds numbers characteristic of ORC turbines, adaptive infill criteria are devised to smartly enrich the surrogate model during optimization on both levels of fidelity. While steady RANS solutions are known to fail in predicting key complex flow features present in turbine vanes, the present surrogate captures elements of the trends from the LES high-fidelity data and provides improved estimations of the better performing designs geometry parameters.

Keywords Multi-fidelity optimization · Organic rankine cycle · Turbine vane · Large eddy simulation

1 Introduction

Advanced aerodynamic shape optimization tools have become increasingly capable of producing efficient turbine vane geometries, and have recently been applied to the Organic Rankine Cycle (ORC) framework [1, 2]. Because the objective function computation usually requires costly flow simulations, especially in complex geome-

C. Matar (✉) · P. Cinnella
Sorbonne University, Paris, France
e-mail: camille.matar@etu.sorbonne-universite.fr

P. Cinnella
e-mail: paola.cinnella@sorbonne-universite.fr

X. Gloerfelt
Arts et Métiers ParisTech, Paris, France
e-mail: xavier.gloerfelt@ensam.eu

A. Giauque et al. (eds.), *Proceedings of the 5th International Seminar on Non-Ideal Compressible Fluid Dynamics for Propulsion and Power*, ERCOFTAC Series 32,
https://doi.org/10.1007/978-3-031-92695-2_9

tries, surrogate models have proved to be good candidates to the overall cost reduction of the design process. Among the surrogate strategies available in the literature, much attention has been given to Kriging models, and various applications (including ORC turbine vane design) have shown its capabilities in providing good estimates of the better performing designs [3]. Both single- and multi-objective global optimizations of blade shape have been performed [4, 5], including robust optimization under uncertainty [6–8]. However, RANS models fail to accurately reproduce several key flow mechanisms that play a major role in turbine loss generation, such as the overall momentum deficit due to the boundary layers or the base pressure behind the trailing edge (TE) [9]. This has also shown to be true for ORC turbines, especially at operating conditions where laminar/turbulent transition, shock wave/boundary layer interactions and wake unsteadiness play an important role.

Unfortunately, the cost of higher-fidelity, turbulent-scale-resolving methods such as Large Eddy Simulation (LES) or hybrid RANS/LES is too prohibitive to be used to compute the objective functions for many designs, especially at the very high Reynolds numbers characteristic of ORC turbines, which has essentially limited ORC design (and of turbomachinery in general) to 2D or 3D RANS simulations only. However, the resulting designs are optimal under the RANS equations at best, with potential misrepresentation of the actual performance. In the effort of seeking for LES-quality optimal designs while reducing the overall computational cost to an amenable budget, we investigate here a multifidelity (MF) approach where only a few, well-chosen designs are simulated with the high-fidelity (HF) LES approach, while the rest of the design space is explored by the low-fidelity (LF) RANS model. For that purpose, we follow the autoregressive co-Kriging procedure described in [10] to generate a nested MF co-Kriging model, which makes the assumption that the RANS and LES predictions are sufficiently well correlated. This is likely not true throughout the design space, but may become more likely as more optimal designs are discovered. Thus, we devise a very few-shots strategy for correcting the LF-based kriging with LES of designs with high probability of improving the optimization criteria.

2 Methodology

2.1 Vane Geometry

The vane geometry was initially designed by [11] using a modified Method of Characteristics (MoC) to account for the non-ideal behavior of the gas during the supersonic expansion, and is shown in Fig. 2a. The Reynolds number based on the blade axial chord $C = 10$ mm and the isentropic exit velocity is $Re_C = 1.3 \times 10^6$, and the target isentropic Mach number downstream of the blade is $M_{2,is} = 1.70$. The working fluid is the refrigerant R134a, often adopted in ORC cycles. The inlet and outlet thermodynamic conditions, along with the target outflow angle β_2 and imposed pressure

Table 1 Vane operating conditions with R134a

Re_C	$M_{2,is}$	Π	$P_{0,1}$[bar]	$T_{0,1}$[K]	β_2[°]	γ	Γ
1.3×10^6	1.70	4.38	4.83	293	75	1.10	0.95

ratio $\Pi = P_{0,1}/P_2$ of the gas ($P_{0,1}$ and P_2 being the inlet stagnation and exit static pressures, respectively) are summarized in Table 1. The fundamental derivative of gas dynamics $\Gamma < 1$ indicates that the thermodynamic conditions of the R134a are located in the so-called dense gas region. This yields opposite variations of the speed of sound across isentropic transformations compared to a perfect gas [12].

All computations are performed with the in-house code MUSICAA, which discretizes the compressible Navier-Stokes equations with high-order finite differences on structured grids (further details can be found in [13]). The LES rely on explicit selective filtering of the unresolved scales (implicit LES). RANS simulations make use of the Spalart-Allmaras one-equation model, following the implementation of [14] to improve numerical robustness. To highlight the large discrepancies between LES and RANS on the baseline geometry, we show in Fig. 1a the boundary layer state on the blade suction side. The shape factor H and friction coefficients C_f strongly disagree on most of the blade suction side, as the RANS predicts a fully turbulent boundary layer starting from the blade leading edge, as evidenced by the lower H and higher C_f. Furthermore, the shockwave/boundary layer interaction (SWBLI) located at $x/C \in [0.75; 0.85]$ is much wider in the case of the LES, as a plateau of negative C_f exists. As the state of the boundary layer impacts both the SWBLI and the flow at the TE, the wake and shock waves downstream of the blade can differ greatly between the two computations. This is shown in Fig. 1b, where the profile of Mach number deviation ΔM from the mixed-out state ($\overline{M} = 1.615$; 1.630 in the RANS and LES, respectively) has been extracted along the line shown in Fig. 2b. As expected, the profiles feature large deviations and the mixed-out values differ by about 1%. Finally, we give in Fig. 2b the mean entropy fields and Mach number contours obtained with both methods. The wake features a more pronounced entropy content in the steady RANS computation, and is slightly wider. Furthermore, the Mach contours near the SWBLI showcase more then one reflected shock wave in the LES compared to the RANS. Therefore, with the overestimated irreversible mixing in the wake and the disparities in Mach number distribution near the outlet, we expect large deviations in blade performance predictions between the two models.

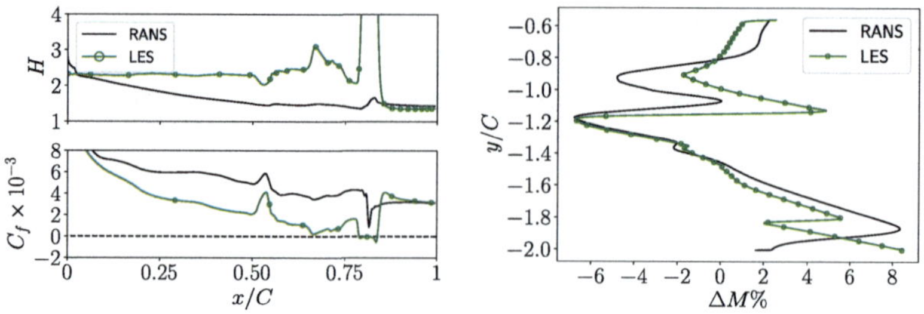

a) Shape factor and skin friction

b) Mean Mach number deviation profile

Fig. 1 **a** Boundary layer shape factor H and friction coefficient C_f on blade suction side, and **b** mean Mach number M deviation profiles in the wake

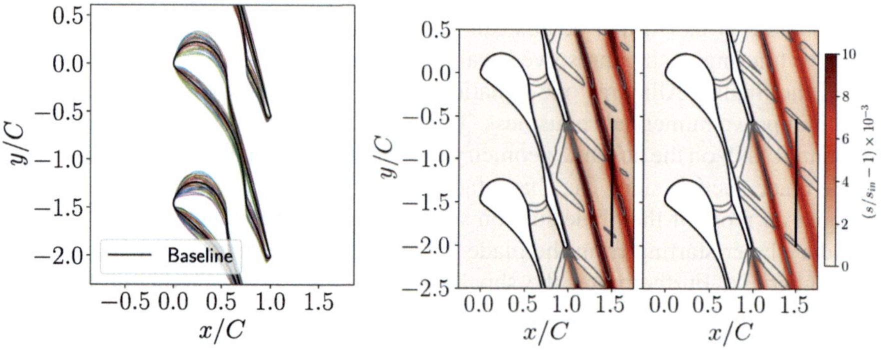

a) Baseline and DoE designs.

b) Left: RANS and right: LES.

Fig. 2 **a** Blade baseline and LF DoE designs and **b** RANS and LES entropy fields with Mach number contours in the range [1; 2], and mixed-out state measurement line

2.2 Optimization Problem

The constrained optimization problem is defined as follows:

$$\text{minimize} \begin{cases} J_1 = \dfrac{\overline{s}}{s_1} \\ J_2 = \sqrt{\dfrac{\sum_{n=1}^{N}\left(\overline{M} - M_n\right)^2}{N}} \end{cases} \quad \text{with} \quad \frac{A_{\text{new}}}{A_{\text{baseline}}} \geq 1 \tag{1}$$

where J_1 is the increase in entropy through the vane, J_2 measures the Mach number uniformity behind the blade and is defined as the standard deviation of the local Mach number over the blade pitch M from a mixed-out state $\overline{M}$, the overbar denoting a mixed-out variable. The constraint is based on the conservation of the initial blade cross-sectional area A_{baseline}. The well established multi-objective Genetic Algorithm

(GA) NSGA-II [15] is used to solve the optimization problem. The blade shape is initially parametrized using Free-Form Deformation (FFD) on 8 control points, and a large number of random designs is generated. Then, a Proper Orthogonal Decomposition (POD) over the geometry database [16] is used to reduce the design space dimentionality, and only 4 geometric modes are deemed sufficient to reproduce all random designs. Thus, the 4 singular values associated to each mode constitute the parameter space. The crossover rate is set to 0.9 and the mutation rate is specified individually for each singular value. Indeed, the first mode carries over 70% of the total modes energy and is thus considered as driving the design space exploration. Therefore, a high mutation rate of 0.2 is associated to it, while a conservative value of 0.1 is attributed to the remaining 3.

2.3 Co-Kriging Model

To model the objective functions response in the parameter space, we use the autoregressive co-Kriging model (also called hierarchical Kriging [17]) formulation proposed in [10] for $s = 2$ levels of code:

$$y_{hf}(\mathbf{x}) = \rho_{lf}(\mathbf{x})y_{lf}(\mathbf{x}) + \mathbf{r}^T\mathbf{R}^{-1}(\mathbf{y}_{S,hf} - \rho_{lf}(\mathbf{x})\mathbf{F}) \tag{2}$$

where $\mathbf{x}$ is the design parameter vector, subscripts hf and lf denote the high- and low-fidelity levels, $\rho_{lf}(\mathbf{x})$ is an adjustment coefficient, $y_{lf}(\mathbf{x})$ is the LF Kriging model prediction, $\mathbf{r}$ is the correlation vector between the untried point and the HF samples, $\mathbf{R}$ the correlation matrix of the HF samples, $\mathbf{y}_{S,hf}$ is the vector of HF samples, and $\mathbf{F}$ is the vector of LF samples at HF locations. Due to the reduced number of high-fidelity samples available, we assume $\rho_{lf}(\mathbf{x}) = \rho_{lf}$ is a constant and that the low- and high-fidelity samples are sufficiently correlated. Furthermore, a Gaussian kernel is chosen for the correlation function. Finally, independent Kriging models are developed for each objective function.

2.4 Infill Strategy

The formalism of Eq. 2 provides direct access to the lower-fidelity model mean y_{lf} and variance σ_{lf}. The latter are particularly useful in estimating optimal additional sample points to be added to the model training dataset during the optimization process. Because the LF function is much cheaper to evaluate, many more samples are available, which renders the LF surrogate more capable of accurately predicting optimal infills. In this work, three additional samples are computed in parallel at every infill and are based on the LF Kriging model. One corresponds to the maximal Probability of Improvement (PI) criterion set to augment the Pareto front [18, 19], and the other two are the minimized single-objective Lower Confidence Bounds (LCB) computed for each objective function $\text{LCB}(\mathbf{x}) = y_{lf}(\mathbf{x}) - A\sigma_{lf}(\mathbf{x})$, where

$A = 1$. In addition to the LF infills, one HF sample is added to the training database every prescribed number of iterations. Because of the nested property of the present co-Kriging model, we simply choose to simulate with the HF method the PI infill chosen by the LF. The choice of frequency for this infill is discussed in the next section, as it is related to the overall computational budget available.

2.5 *Validation and Budget*

2.5.1 Single-Fidelity Kriging

To validate the described methodology, we first converge the optimization without a Kriging model solely using RANS on a coarse mesh, denoted LF RANS. With an initial population size of 30 individuals, 70 generations are required to converge the procedure, resulting in a total of 2 100 function evaluations with the flow solver. Each LF RANS costs 4.8CPUh on 57 CPUs (31 875 points in the mesh) and the optimization is run on a supercomputer, thus a total of 10 080CPUh are required. The resulting Pareto front is shown in Fig. 3a. It is well distributed and features a discontinuity near the high J_1 and low J_2 limit, as a result of the constraint. This shows that the direct optimization procedure is clearly capable of exploring a wide design space. Then, we perform the optimization with a Kriging model for the LF RANS only. The Design of Experiments (DoE) size is $10 \times N_p = 40$, where $N_p = 4$ is the number of design parameters. The DoE designs are chosen with a latin hypercube sampling of the parameter space, and are shown in Fig. 2a and in Fig. 3b in the functional space. Then, the optimization is enriched with infills as described in Sect. 2.4, and a summary of these parameters is given in Table 2. After extensive testing, we find that an infill frequency of once every 20 generations is sufficient, and a total of 5 infill calls is made, amounting to 100 generations. Then, the optimization is further converged with the Kriging model only for another 50 generations, and the resulting Pareto front is shown in Fig. 3a. To validate the Kriging predictions, we select optimal designs spread evenly along the front and simulate them with RANS (Kriging recomp RANS LF). The agreement is good as the surrogate model successfully predicts true Pareto optimal solutions, albeit the aforementioned discontinuity is not resolved.

2.5.2 MF Co-Kriging

While the number of LF RANS samples can be large, the LES observations are computationally expensive and are thus very limited. In practice, a finite computational resource budget is available, and a compromise between the number of HF infills and DoE samples must be made. For this study, the total available budget was 1 000 000CPUh. The cost of a single LES with the present structured solver is $\approx$ 130 000CPUh (440M points in the mesh, 4 704 CPUs), thus only eight LES can be run, of which one must be kept for validating at least one Pareto optimal design,

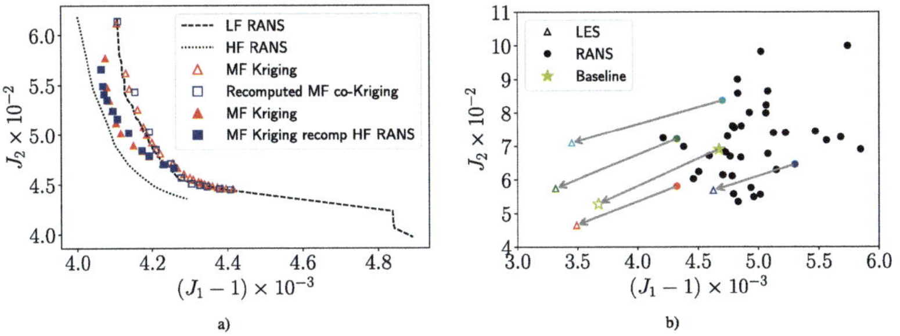

Fig. 3 **a** Pareto fronts obtained with both LF and HF RANS simulations, the Kriging model, and the MF co-Kriging model. **b** RANS Design of Experiments (DoE) in the objective function space, with recomputed candidates via LES

Table 2 Summary of optimization parameters. N_d is the number of designs per generation, N_g the number of generations, $N_{lf,i}$ and $N_{hf,i}$ the numbers of LF and HF infills, $N_{lf,total}$ and $N_{hf,total}$ the number of total LF and HF methods calls

Run	N_d	N_g	$N_{lf,i}$ ($\times 3$)	$N_{hf,i}$	Last infill generation	$N_{lf,total}$	$N_{hf,total}$
LF RANS	30	70	–	–	–	2 100	–
Kriging	30	150	5	–	100	56	–
MF co-Kriging	30	150	5	2	100	56	7

and another is used to simulate the baseline configuration, leaving six LES to be distributed among the DoE and infills. Usually, a latin hyper-cube sampling of the parameter space would be used to find appropriate HF samples, and corresponding LF points would be added to the DoE (to respect the nested property of the co-Kriging model). However, with the very small number of LES available and the high risk associated with simulating low performing designs that the optimization procedure could rapidly discard, we build the HF DoE with Pareto optimal and extreme (low J_1 & high J_2 and vice versa) designs, plus the baseline configuration. A series of tests using just RANS on two grid levels where the finer grid is refined 2 times in each direction, leading to a LF RANS and a HF RANS, indicates that 4 designs in addition to the baseline in the DoE, and 2 infills every other LF infill yields promising results. The MF co-Kriging Pareto front is given in Fig. 3a, and candidates have been recomputed using the HF RANS. Furthermore, Pareto optimal candidates from the LF RANS were also recomputed with the HF RANS. The overall trend is satisfactory, considering the number of total function evaluations with the HF RANS of only 7 (summarized in Table 2) and just 56 with the LF RANS. The co-Kriging correctly predicts the best compromise designs, but fails in the low J_2 region, corresponding to vanes where the Mach number uniformity is strong.

3 Results

3.1 Pareto Front Analysis

The performance of the LES DoE designs are shown in Fig. 3b. The arrows indicate the RANS predictions of the corresponding designs. Evidently, the LF computations largely overestimate J_1 and to a lesser extent J_2, which was anticipated in the observations made in Sect. 2.1. Nonetheless, we observe a reasonable correlation between both levels of fidelity which initially supports the Kriging model founding hypothesis, as well as the present assumption of constant adjustment coefficient $\rho_{lf}(\mathbf{x}) = \rho_{lf}$. Figure 4 shows the LES Pareto front predicted with the MF co-Kriging model, along with the 2 infills evaluated during the optimization procedure. The PI criterion appears to have efficiently selected augmenting designs to re-train the surrogate in the region near the Pareto front. While this allows to quickly exploit the DoE, the exploration of the design space is subsequently limited and interesting designs may have been overlooked. A Pareto optimal design was then selected (Selected) and recomputed with the last LES available (Recomp selected), and its true performance are given in the same figure. Good agreement between the surrogate prediction and the true performance is achieved considering the overall low number of HF and LF calls, resulting in deviations of $\approx 5.6\%$ and $\approx 2.4\%$ in $J_1 - 1$ and J_2 objective functions, respectively. Comparatively, the HF RANS Pareto designs have an overall minimum error with respect to the selected LES of $\approx 24.6\%$ and $\approx 12.7\%$ in $J_1 - 1$ and J_2, respectively, translating into a reduction in error by factors of about ≈ 4.4 and ≈ 5.3.

Interestingly, the MF co-Kriging surrogate based on LF and HF RANS computations appears to perform better at predicting optimal designs of the HF RANS, when compared to the surrogate used for the LES. Since both optimizations rely on the same DoE size and quantity of infills, this hints that the number of HF samples (be they RANS or LES) is not the sole driving factor behind the performance of the MF surrogate. Rather, the correlation between the LF and HF samples is the most dominant parameter, which in the co-Kriging model is encapsulated within the adjustment

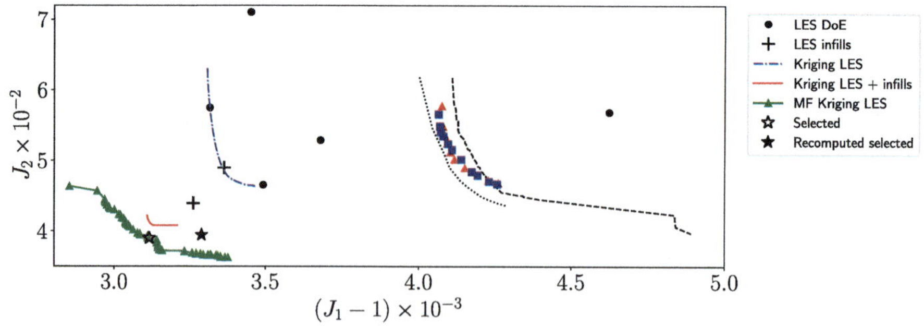

Fig. 4 Comparison of predicted Pareto fronts from several optimizations

coefficient $\rho_{lf}(\mathbf{x})$. In this view, increasing the DoE size allows to compensate for a poor correlation between two levels of fidelity, while fewer samples are required to efficiently train a MF surrogate on highly correlated data. On the one hand, the latter case was verified in the MF RANS optimization as the LF and HF RANS predictions are very well correlated, on account of the fidelity being quantified by the level of grid refinement which only seemed to slightly alter the solution in the present numerical work. There, assuming $\rho_{lf}(\mathbf{x}) = \rho_{lf}$ is relevant as only a simple scaling is required to correct most of the LF RANS predictions. On the other hand, the *a priori* strong correlation between the RANS and LES DoE observed in Fig. 3b may have in fact deteriorated as the optimization procedure reached the Pareto front region. This suggests that the present RANS methodology is capable of capturing the global trends of blade performance variation with large shape deformations, while it fails at correctly predicting the impact of fine geometric perturbations on blade losses. Furthermore, while the test series on the LF and HF RANS allowed to estimate the optimization and Kriging model parameters, they are not truly representative of the practical issues encountered when dealing with both low-cost steady computations and expensive scale-resolving simulations.

In the MF co-Kriging surrogate, the LF RANS serves to explore the design space at a fraction of the cost of the LES, and its predictions are subsequently corrected by the model in Eq. 2. However, if the correlation between the RANS and LES is weak and the amount of data to allow for more complex forms of the adjustment coefficient $\rho_{lf}(\mathbf{x})$ (linear or quadratic, rather than constant) is insufficient, one could suspect the MF co-Kriging approach to under perform when compared to a single-fidelity model solely based on the very few LES samples. This point is inspected in the following. We build two Kriging surrogates: one (Kriging LES) trained only on the 5 LES samples constituting the original DoE, and another (Kriging LES + infills) trained on the same DoE augmented by the two infills computed during the MF co-Kriging optimization. We note that the second Kriging surrogate not only beneficiates from a 40% larger database but also from the two infills having been selected to promote exploitation of the design space, thus giving it an additional advantage over the Kriging surrogate trained on the smaller DoE. The resulting Pareto fronts are provided in Fig. 4. Clearly, 5 samples is insufficient to train the Kriging surrogate (Kriging LES) and the Pareto front provides virtually no improvement over the DoE designs. Upon adding the two infills, the Kriging surrogate (Kriging LES + infills) approaches the MF co-Kriging Pareto front, although it only covers a very small region of the functional space. This is most likely due to the infills being chosen in the optimal designs region rather than from a well-distributed sampling of the design space. In addition, the LES + infills Pareto front is dominated by the "recomputed" points (ground truth LES), as a result of the co-Kriging surrogate inaccuracy when being trained on a very reduced dataset.

3.2 RANS and LES Designs Comparison

We now compare 3 Pareto optimal designs predicted by the MF co-Kriging surrogates for both LES and RANS optimizations in Fig. 5a, consisting in the minimum of each objective function and the most centered along the front. Interestingly, the correlation of the increasing deformation of the LE and thinning TE with the decrease (increase) in J_1 (J_2) is similar between both levels of fidelity. Furthermore, this trend indicates that the Mach number field downstream of the blade becomes more uniform with thicker TE, whereas the associated entropy generation is enhanced. We note however that the overall deformations are milder in the LES optimal designs, and this is discussed in the following. We report in Fig. 5b the streamwise velocity field in the LE vicinity inside a RANS Pareto optimal (Left) and the LES DoE worst (Right) designs. We note that both geometries are deformed in a similar fashion and feature a strong curvature on the pressure side. This causes boundary layer separation leading to a large recirculation bubble in the RANS, and to a lesser extent in the LES. Surprisingly, this is a common trait shared by all RANS Pareto optimal candidates,

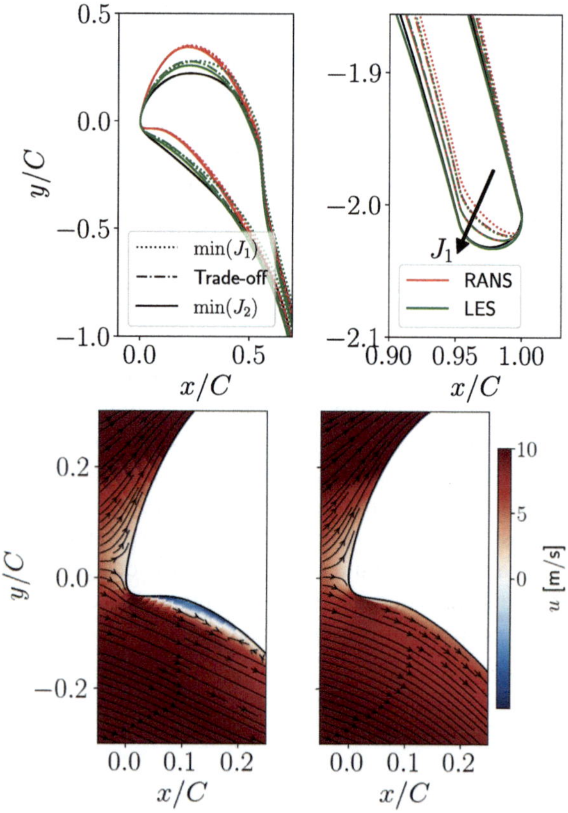

Fig. 5 **a** Pareto optimal geometries predicted by both RANS and LES MF co-Kriging surrogates, **b** streamwise velocity field inside (Left) RANS Pareto optimal and (Right) LES worst DoE geometries

which indicates that the additional entropy generated by the flow separation and recirculation only plays a minor role in the performance reduction of the blade when computed with the present RANS. This is made possible by the assumption of fully turbulent boundary layer which causes global losses largely surpassing that of the LE separation. On the other hand, the much smaller separated flow region in the LES worst design is sufficient to drastically reduce the performance of the blade compared to other candidates computed with LES. This point is elaborated further in the next section.

3.3 Role of LE and TE in the LES

To illustrate the role of the LE and TE topology on blade performance as predicted by the LES, we compare three geometries: the Selected (which served to validate the MF co-Kriging model, see Fig. 4), the Baseline, and the worst LES DoE one, from hereon denoted as LE1. We note that these candidates dominate each other in that respective order. The three geometries are provided in Fig. 6a. First, we compare the wake entropy profiles at several locations downstream of the blades, given in Fig. 6b. The widest wake is obtained in the Baseline design which is attributable to its TE being the thickest. Then, the strongest entropy content is reached in the LE1 design, which also produces the thinnest wake. Interestingly, the Selected design produces a slightly thicker wake compared to the LE1 and yet generates the less entropy overall. Therefore, the maximum entropy content inside the wake does not vary monotonically with TE thickness. In fact, it can be shown that the LE plays a major role in the performance of the blade.

As discussed in an earlier section, the highly deformed LE of the LE1 design causes flow separation on the pressure side. To investigate the impact on the blade performance in detail, we compute the pressure side boundary layer momentum thickness θ as well as dissipation coefficient [20], which we split into its steady and unsteady contributions $C_d = \mu\Phi + \rho\epsilon$. The results are given in Fig. 7. In the LE1 geometry, θ rapidly increases in the LE region compared to the Baseline and Selected designs, and decreases inside the vane due to the strong favorable pressure gradient. A similar behavior is assumed by the unsteady contribution $\rho\epsilon$, which testifies that flow structures are being shed from the LE only in the LE1 geometry, enhancing kinetic energy dissipation. Furthermore, both $\mu\Phi$ and $\rho\epsilon$ sharply increase just upstream of the TE near $x/C = 0.9$ as a result of the initially perturbed boundary layer near the LE, showing the strong history effects inside the boundary layer. This finally results in additional kinetic energy losses compared to the Baseline and Selected geometries.

As expected, these complex flow phenomena are not resolved nor modeled by the present RANS strategy. Therefore, the optimization procedure relying on such strong modeling assumptions and misrepresentations of loss generating mechanisms guides the algorithm towards erroneous design solutions.

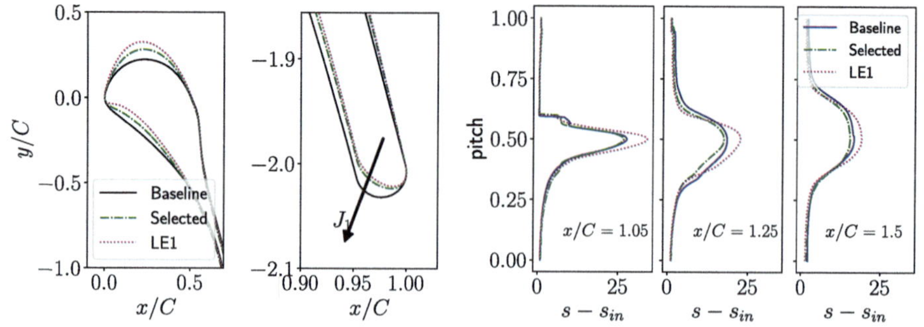

Fig. 6 **a** LES designs details around the LE and TE, **b** entropy deviation profiles across the LES wakes at three locations downstream of the blades

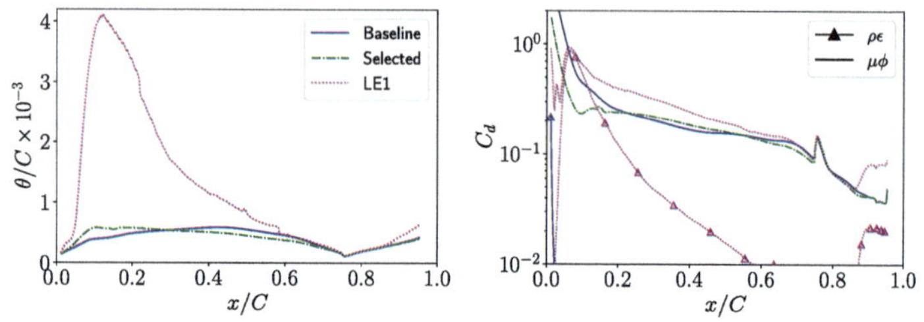

Fig. 7 Pressure side: **a** momentum thickness, **b** dissipation coefficient C_d

4 Conclusions

In this work, we performed shape optimization of a high-pressure turbine vane using a multi-fidelity (MF) co-Kriging surrogate of both steady RANS computations and wall-resolved LES. The methods for generating an appropriate Design of Experiments for both levels is described, along with the infill strategy. The MF approach is validated in a test case using RANS only where the high-fidelity (HF) objective function evaluation consists in a fine mesh computation. This allows to estimate optimal co-Kriging parameters, such as the frequency of low-fidelity (LF) and HF infills and the number of generations required to reach convergence. A total of 8 LES are performed to train the MF co-Kriging, while 56 LF samples are generated. The surrogate successfully captures the correlation between the LES and RANS observations, and provides a good approximation of the better performing designs. On the contrary, a surrogate-based optimization using only the small set of LES samples does not converge toward non-dominated designs. Subsequently, the geometries constituting the RANS and LES Pareto fronts are compared, and highlight major differences in performance predictions. Specifically, the RANS provides designs with highly deformed LE and TE compared to the LES, which triggers flow separation near the LE and is highly detrimental only in the LES. This stems from the RANS boundary layers being fully turbulent, thus contributing to entropy production in much larger proportions compared to the LE flow separation, of which the relative contribution to losses is rendered negligible. Overall, the need for scale-resolved data in the shape optimization loop is demonstrated, as well as the capacity of a MF co-Kriging model to account for large deviations in objective function predictions.

References

1. Al Jubori, A., Al-Dadah, R.K., Mahmoud, S., Bahr Ennil, A.S., Rahbar, K.: Three dimensional optimization of small-scale axial turbine for low temperature heat source driven organic Rankine cycle. Energy Conver. Manage. **133**(1), 411–426 (2017)
2. Witanowski, L., Klonowicz, P., Lampart, P., Suchocki, T., Jedezejewski, L., Zanievski, D.: Optimization of an axial turbine for a small scale ORC waste heat recovery system. Energy **205**(1), 1–18 (2020)
3. Rodriguez-Fernandez, P., Persico, G.: Automatic design of ORC turbine profiles using evolutionary algorithms. In: 3rd International Seminar on ORC Power Systems, pp. 1–10 (2015)
4. Persico, G., Rodriguez-Fernandez, P., Romei, A.: High-fidelity shape optimization of non-conventional turbomachinery by surrogate evolutionary strategies. J. Turbomach. **141**(18), 1–11 (2019)
5. Espinosa Sarmiento, A.L., Ramirez Camacho, R.G., de Olibeira, W., Gutiérrez Velasquez, E.I., Murthi, M., Diaz Gautier, N.L.: Design and off-design performance improvement of a radial-inflow turbine for ORC applications using metamodels and genetic algorithm optimization. Appl. Thermal Eng. **183**(1), 1–14 (2021)
6. Congedo, P.M., Corre, C., Martinez, J.-M.: Shape optimization of an airfoil in a BZT flow with multiple-source uncertainties. Comput. Methods Appl. Mech. Eng. **200**(1), 216–232 (2011)
7. Bufi, E.A., Cinnella, P.: Robust optimization of supersonic ORC nozzle guide vanes. In: Journal of Physics: Conference Series, vol. 821, no. 1, p. 012014. IOP Publishing (2017)
8. Serafino, A., Obert, B., Cinnella, P.: Multi-fidelity robust design optimization of an ORC turbine for high temperature waste heat recovery. Energy **269**(1), 1–13 (2023)
9. Sandberg, R.D., Michelassi, V.: Fluid dynamics of axial turbomachinery: blade- and stage-level simulations and models. Ann. Rev. Fluid Mech. **54**(1), 255–285 (2022)
10. Le Gratiet, L., Garnier, J.: Recursive co-Kriging model for design of computer experiments with multiple levels of fidelity. Int. J. Uncertainty Quant. **4**(5), 365–386 (2014)
11. David, B., John, J.O., Andrew, P.S.W.: The effect of Isentropic exponent on transonic turbine performance. J. Turbomach. **142**(1) (2020)
12. Thompson, P.A.: A fundamental derivative in gas dynamics. Phys. Fluids **14**(9), 1843–1849 (1971)
13. Bienner, A., Gloerfelt, X., Yalçin, Ö., Cinnella, P.: Multiblock parallel high-order implicit residual smoothing time scheme for compressible Navier-Stokes equations. Comput. Fluids **269**(1), 1–25 (2024)
14. Crivellini, A., D'Alessandro, V., Bassi, F.: A Spalart-Allmaras turbulence model implementation in a discontinuous Galerkin solver for incompressible flows. J. Comput. Phys. **241**(1), 388–415 (2019)
15. Deb, K., Pratap, A., Agarwal, S., Meyarivan, T.: A fast and elitist multiobjective genetic algorithm: NSGA-II. IEEE Trans. Evol. Comput. **6**(2), 182–197 (2002)
16. Toal, D.J.J., Bressloff, N.W., Keane, A.J.: Geometric filtration using POD for aerodynamic design optimization. In: 26th AIAA Applied Aerodynamics Conference, pp. 1–13 (2008)
17. Han, Z.-H., Zimmermann, R., Görtz, S.: Alternative cokriging model for variable-fidelity surrogate modeling. AIAA J. **50**(5), 1–6 (2012)

18. Keane, A.J.: Statistical improvement criteria for use in multiobjective design optimization. AIAA J. **44**(4), 1–13 (2006)
19. Zhang, Y., Han, Z.-H., Zhang, K.-S.: Variable-fidelity expected improvement method for efficient global optimization of expensive functions. Struct. Multidiscip. Optim. **58**(1), 1431–1451 (2018)
20. Ghasemi, E., McEligot, D.M., Nolan, K.P., Crepeau, J., Siahpush, A., Budwig, R.S., Tokuhiro, A.: Effects of adverse and favorable pressure gradients on entropy generation in a transitional boundary layer region under the influence of freestream turbulence. Int. J. Heat Mass Transfer **77**, 475–488 (2014)

Thermodynamically Consistent Interpolation for Tabulated Equations of State

Paolo Errante and Christophe Corre

Abstract Computational fluid dynamics (CFD) simulations involving non-ideal gases or gas mixtures rely on thermodynamic models which can be computationally expensive. Look-Up Tables (LUTs) have been extensively used to reduce such computational costs. During the simulation, the CFD solver accesses pre-computed discrete values of thermodynamic properties which are stored in a LUT. Interpolation techniques are then essential to accurately estimate the fluid local properties for any intermediate state between those available in the LUT. The interpolation technique employed needs to guarantee thermodynamic consistency, meaning the interpolated properties satisfy the fundamental laws of thermodynamics. To reconstruct an interpolation satisfying these constraints, bi-quintic Hermite polynomials have been proposed by Swesty [1], which are built from a thermodynamic potential and its successive derivatives. The same technique was successfully used by Zeman et al. [2] for laser plasma investigation. The present paper successfully applies the bi-quintic interpolation in the context of NICFD simulations and establishes its superior accuracy with respect to conventional bilinear interpolation.

Keywords Look-up tables · Thermodynamic consistency · Bi-quintic hermite interpolation · Compact high-order finite-difference formulae

1 Introduction

The accurate description of thermodynamic properties is crucial for Non-Ideal Compressible Fluid Dynamics (NICFD). The predictive numerical simulation of non-ideal gas flows relies heavily on precise estimates of thermodynamic properties such as speed of sound or fundamental derivative. A review of the strengths and limita-

P. Errante (✉)
Lille Fluid Mechanics Laboratory (LMFL), Lille, France
e-mail: paolo.errante@ensam.eu

C. Corre
Fluid Mechanics and Acoustics Laboratory (LMFA), Lyon, France
e-mail: christophe.corre@ec-lyon.fr

A. Giauque et al. (eds.), *Proceedings of the 5th International Seminar on Non-Ideal Compressible Fluid Dynamics for Propulsion and Power*, ERCOFTAC Series 32,
https://doi.org/10.1007/978-3-031-92695-2_10

tions of various families of Equations of State (EoS) expressing relations between thermodynamic properties may be found in the recent review paper [3] devoted to NICFD of dense vapors and supercritical fluids. The point of interest for the present work is that many Computational Fluid Dynamics (CFD) softwares rely on external libraries for EoS calculations. These routines often evaluate thermodynamic properties using independent variables different from those involved in the flow governing equations, implying intermediate projections of the state from the EoS library to the CFD solver with increased costs and potential lack of efficiency [4]. A powerful tool employed to provide accurate thermodynamic description of Non-Ideal gases is Molecular Dynamic (MD) simulation. MD offers a detailed picture of the fluid at atomic levels, allowing observation of phenomena such as diffusion and chemical reactions and represents a direct method to compute macroscopic properties from classical statistical thermodynamic. However, simulating large systems or long time scales becomes computationally prohibitive for fluid dynamics simulations. To overcome the drawbacks of these approaches, LUTs represent a valuable alternative to reduce computational costs. This technique consists in pre-computing thermodynamic properties using one of the methods described above and feeding the values to a flow solver during the simulation. As LUTs only contain pre-calculated properties with respect to an arbitrary set of independent thermodynamic coordinates, interpolation techniques are essential to estimate local properties for any intermediate state encountered during a simulation. An appropriate choice of the interpolating function employed with LUTs is critical not only for numerical accuracy but also for numerical stability. A naive way to tackle the interpolation problem is to make use of independent interpolation functions for each of the properties. Such a strategy would work provided the LUT is sufficiently refined but would also require additional tuning when generating the LUT. A better way to proceed would be to require the interpolation scheme to preserve thermodynamic consistency. In other words, the interpolated quantities should obey fundamental relationships between state variables established by laws of thermodynamics as well as Maxwell relations. This property can be satisfied by choosing an appropriate (at least C^2) function that interpolates a thermodynamic potential from which all other properties can be analytically derived. It is worth pointing out, as done in [3], that a rigorous analysis for the impact of employing consistent or inconsistent interpolations when computing thermodynamic properties for NICFD simulations is still lacking.

Swesty [1] proposed an interpolation scheme based on bi-quintic Hermite polynomials which ensures thermodynamic consistency in the context of astrophysical simulations; the EOS considered as an application dealt with relativistic electron-positron gas. The same ideas were more recently used by [2] for laser plasma investigation performed using Lagrangian and ALE hydrodynamic simulations. The purpose of the present work is to perform a preliminary assessment of such a consistent interpolation in the NICFD context.

Section 2 reviews the consistent biquintic Hermitian interpolation procedure proposed by Swesty [1] and briefly explains how compact and accurate FD estimates can be built from the quantities available in the LUT. Section 3 presents some preliminary results demonstrating high-accuracy is indeed achieved using the Hermitian interpolation.

2 Consistent Lut Interpolation Using Swesty's Approach

2.1 General Principles

Let us assume the LUT contains values of Helmholtz free energy and its first and second derivatives at discrete couples (T_i, ρ_j); at each node in the temperature-density space, the LUT stores $f_{i,j} = f(T_i, \rho_j)$, $(f_T)_{i,j}$, $(f_\rho)_{i,j}$, $(f_{TT})_{i,j}$, $(f_{\rho\rho})_{i,j}$, $(f_{T\rho})_{i,j}$ (see Fig. 1).

Using the available exact values of T_i, ρ_j, $f_{i,j}$, $(f_T)_{i,j}$, $(f_\rho)_{i,j}$, $(f_{TT})_{i,j}$, $(f_{\rho\rho})_{i,j}$, $(f_{T\rho})_{i,j}$, all physical quantities of interest can be computed at the nodes of the LUT: specific entropy is computed as $s = -f_T$, specific internal energy as $e = f - Tf_T$, pressure as $p = \rho^2 f_\rho$, specific isochoric heat capacity as $c_v = -Tf_{TT}$, speed of sound as $a = \sqrt{2\rho f_\rho + \rho^2 f_{\rho\rho} - \rho^2 f_{TT}^{-1} f_{\rho T}^2}$.

Assuming in a first step that ρ is fixed, we wish to build an interpolation to compute f at any temperature between T_i and T_{i+1} such that we recover the known values of f, f_T, f_{TT} at $T = T_i$ and $T = T_{i+1}$. Following Swesty, a quintic Hermitian interpolation is used:

$$\begin{aligned}\tilde{f}(T, \rho_j) &= f_{i,j}\psi_x^0 + f_{i+1,j}\psi_{1-x}^0 + (f_T)_{i,j}\Delta T\psi_x^1 - (f_T)_{i+1,j}\Delta T\psi_{1-x}^1 \\ &\quad +(f_{TT})_{i,j}\Delta T^2\psi_x^2 + (f_{TT})_{i+1,j}\Delta T^2\psi_{1-x}^2\end{aligned}$$

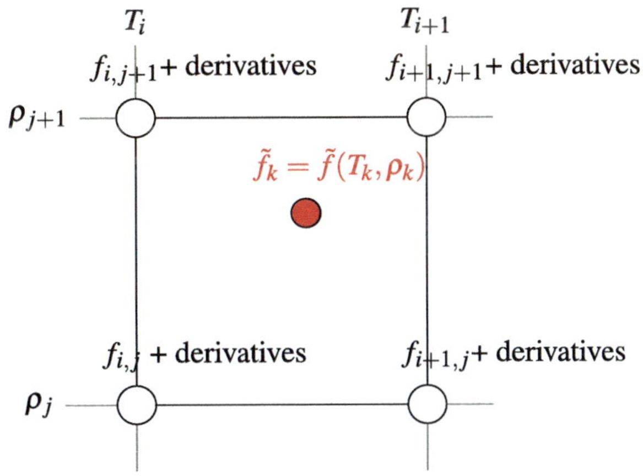

Fig. 1 Typical cell of the LUT for Helmholtz free energy $f(T, \rho)$

where ΔT denotes the LUT discretization step for temperature and the ψ functions are quintic Hermite basis functions defined as follows:

$$\psi_z^0 = \psi^0(z) = -6z^5 + 15z^4 - 10z^3 + 1,\ \psi_z^1 = \psi^1(z) = -3z^5 + 8z^4 - 6z^3 + z,\ \psi_z^2 = \psi^2(z) = \frac{1}{2}(-z^5 + 3z^4 - 3z^3 + z^2)$$

with $x = \dfrac{T - T_i}{\Delta T}$. The polynomial interpolation $\tilde{f}(T, \rho)$ in each cell of the LUT is obtained by "tensorizing" the previous 1D formula. Let us introduce the constant step $\Delta\rho = \rho_{j+1} - \rho_j$ and the reduced variable $y = \dfrac{\rho - \rho_j}{\Delta\rho}$. Let us also make use of the following more compact notations:

$$\bar{f}_T^{i,j} = (f_T)_{i,j}\Delta T,\quad \bar{f}_{TT}^{i,j} = (f_{TT})_{i,j}\Delta T^2,\quad \bar{f}_\rho^{i,j} = (f_\rho)_{i,j}\Delta\rho,\quad \bar{f}_{\rho\rho}^{i,j} = (f_{\rho\rho})_{i,j}\Delta\rho^2,$$
$$\bar{f}_{T\rho}^{i,j} = (f_{T\rho})_{i,j}\Delta T\Delta\rho,\ \bar{f}_{TT\rho}^{i,j} = (f_{TT\rho})_{i,j}\Delta T^2\Delta\rho,\ \bar{f}_{T\rho\rho}^{i,j} = (f_{T\rho\rho})_{i,j}\Delta T\Delta\rho^2,\ \bar{f}_{TT\rho\rho}^{i,j} = (f_{TT\rho\rho})_{i,j}\Delta T^2\Delta\rho^2$$

The bi-quintic Hermitian interpolation then reads:

$$\begin{aligned}
\tilde{f}(T,\rho) = {} & f^{i,j}\psi_x^0\psi_y^0 + f^{i+1,j}\psi_{1-x}^0\psi_y^0 + f^{i,j+1}\psi_x^0\psi_{1-y}^0 + f^{i+1,j+1}\psi_{1-x}^0\psi_{1-y}^0 \\
& + \bar{f}_T^{i,j}\psi_x^1\psi_y^0 - \bar{f}_T^{i+1,j}\psi_{1-x}^1\psi_y^0 + \bar{f}_T^{i,j+1}\psi_x^1\psi_{1-y}^0 - \bar{f}_T^{i+1,j+1}\psi_{1-x}^1\psi_{1-y}^0 \\
& + \bar{f}_{TT}^{i,j}\psi_x^2\psi_y^0 + \bar{f}_{TT}^{i+1,j}\psi_{1-x}^2\psi_y^0 + \bar{f}_{TT}^{i,j+1}\psi_x^2\psi_{1-y}^0 + \bar{f}_{TT}^{i+1,j+1}\psi_{1-x}^2\psi_{1-y}^0 \\
& + \bar{f}_\rho^{i,j}\psi_x^0\psi_y^1 + \bar{f}_\rho^{i+1,j}\psi_{1-x}^0\psi_y^1 - \bar{f}_\rho^{i,j+1}\psi_x^0\psi_{1-y}^1 - \bar{f}_\rho^{i+1,j+1}\psi_{1-x}^0\psi_{1-y}^1 \\
& + \bar{f}_{\rho\rho}^{i,j}\psi_x^0\psi_y^2 + \bar{f}_{\rho\rho}^{i+1,j}\psi_{1-x}^0\psi_y^2 + \bar{f}_{\rho\rho}^{i,j+1}\psi_x^0\psi_{1-y}^2 + \bar{f}_{\rho\rho}^{i+1,j+1}\psi_{1-x}^0\psi_{1-y}^2 \\
& + \bar{f}_{T\rho}^{i,j}\psi_x^1\psi_y^1 - \bar{f}_{T\rho}^{i+1,j}\psi_{1-x}^1\psi_y^1 - \bar{f}_{T\rho}^{i,j+1}\psi_x^1\psi_{1-y}^1 + \bar{f}_{T\rho}^{i+1,j+1}\psi_{1-x}^1\psi_{1-y}^1 \\
& + \bar{f}_{TT\rho}^{i,j}\psi_x^2\psi_y^1 + \bar{f}_{TT\rho}^{i+1,j}\psi_{1-x}^2\psi_y^1 - \bar{f}_{TT\rho}^{i,j+1}\psi_x^2\psi_{1-y}^1 - \bar{f}_{TT\rho}^{i+1,j+1}\psi_{1-x}^2\psi_{1-y}^1 \\
& + \bar{f}_{T\rho\rho}^{i,j}\psi_x^1\psi_y^2 - \bar{f}_{T\rho\rho}^{i+1,j}\psi_{1-x}^1\psi_y^2 + \bar{f}_{T\rho\rho}^{i,j+1}\psi_x^1\psi_{1-y}^2 - \bar{f}_{T\rho\rho}^{i+1,j+1}\psi_{1-x}^1\psi_{1-y}^2 \\
& + \bar{f}_{TT\rho\rho}^{i,j}\psi_x^2\psi_y^2 + \bar{f}_{TT\rho\rho}^{i+1,j}\psi_{1-x}^2\psi_y^2 + \bar{f}_{TT\rho\rho}^{i,j+1}\psi_x^2\psi_{1-y}^2 + \bar{f}_{TT\rho\rho}^{i+1,j+1}\psi_{1-x}^2\psi_{1-y}^2
\end{aligned} \tag{1}$$

where the high-order derivatives (HOD) (third and fourth-order) are not necessarily available in the LUT. Also HOD make the expansion more expensive to compute: however we will show next that HOD are necessary to ensure the accuracy of the interpolation.

All physical quantities can be recovered by immediate differentiation with respect to x and/or y: only the polynomial $\psi_{x/y}^{0/1/2}$ needs to be differentiated, all other terms being constant for the cell under consideration. Let us denote

$$d\psi_z^0 = \frac{d\psi^0}{dz},\ d^2\psi_z^0 = \frac{d^2\psi^0}{dz^2},\ d\psi_z^1 = \frac{d\psi^1}{dz},\ d^2\psi_z^1 = \frac{d^2\psi^1}{dz^2},\ d\psi_z^2 = \frac{d\psi^2}{dz},\ d^2\psi_z^2 = \frac{d^2\psi^2}{dz^2}$$

The pressure polynomial in a LUT cell is obtained from $\tilde{p}(T,\rho) = \rho^2 \left(\frac{\partial \tilde{f}}{\partial \rho}\right)_T$. The interpolation remains consistent because the same polynomial $\tilde{f}$ is systematically used to compute all thermodynamic quantities from the Helmholtz free energy. For pressure interpolation, we obtain:

$$
\begin{aligned}
\tilde{p}(T,\rho) = \rho^2 (& f^{i,j}\psi_x^0 d\psi_y^0 + f^{i+1,j}\psi_{1-x}^0 d\psi_y^0 - f^{i,j+1}\psi_x^0 d\psi_{1-y}^0 - f^{i+1,j+1}\psi_{1-x}^0 d\psi_{1-y}^0 \\
& + \bar{f}_T^{i,j}\psi_x^1 d\psi_y^0 - \bar{f}_T^{i+1,j}\psi_{1-x}^1 d\psi_y^0 - \bar{f}_T^{i,j+1}\psi_x^1 d\psi_{1-y}^0 + \bar{f}_T^{i+1,j+1}\psi_{1-x}^1 d\psi_{1-y}^0 \\
& + \bar{f}_{TT}^{i,j}\psi_x^2 d\psi_y^0 + \bar{f}_{TT}^{i+1,j}\psi_{1-x}^2 d\psi_y^0 - \bar{f}_{TT}^{i,j+1}\psi_x^2 d\psi_{1-y}^0 - \bar{f}_{TT}^{i+1,j+1}\psi_{1-x}^2 d\psi_{1-y}^0 \\
& + \bar{f}_\rho^{i,j}\psi_x^0 d\psi_y^1 + \bar{f}_\rho^{i+1,j}\psi_{1-x}^0 d\psi_y^1 + \bar{f}_\rho^{i,j+1}\psi_x^0 \psi_{1-y}^1 + \bar{f}_\rho^{i+1,j+1}\psi_{1-x}^0 d\psi_{1-y}^1 \\
& + \bar{f}_{\rho\rho}^{i,j}\psi_x^0 d\psi_y^2 + \bar{f}_{\rho\rho}^{i+1,j}\psi_{1-x}^0 d\psi_y^2 - \bar{f}_{\rho\rho}^{i,j+1}\psi_x^0 d\psi_{1-y}^2 - \bar{f}_{\rho\rho}^{i+1,j+1}\psi_{1-x}^0 d\psi_{1-y}^2 \\
& + \bar{f}_{T\rho}^{i,j}\psi_x^1 d\psi_y^1 - \bar{f}_{T\rho}^{i+1,j}\psi_{1-x}^1 d\psi_y^1 + \bar{f}_{T\rho}^{i,j+1}\psi_x^1 d\psi_{1-y}^1 - \bar{f}_{T\rho}^{i+1,j+1}\psi_{1-x}^1 d\psi_{1-y}^1 \\
& + \bar{f}_{TT\rho}^{i,j}\psi_x^2 d\psi_y^1 + \bar{f}_{TT\rho}^{i+1,j}\psi_{1-x}^2 d\psi_y^1 + \bar{f}_{TT\rho}^{i,j+1}\psi_x^2 \psi_{1-y}^1 + \bar{f}_{TT\rho}^{i+1,j+1}\psi_{1-x}^2 d\psi_{1-y}^1 \\
& + \bar{f}_{T\rho\rho}^{i,j}\psi_x^1 d\psi_y^2 - \bar{f}_{T\rho\rho}^{i+1,j}\psi_{1-x}^1 d\psi_y^2 - \bar{f}_{T\rho\rho}^{i,j+1}\psi_x^1 d\psi_{1-y}^2 + \bar{f}_{T\rho\rho}^{i+1,j+1}\psi_{1-x}^1 d\psi_{1-y}^2 \\
& + \bar{f}_{TT\rho\rho}^{i,j}\psi_x^2 d\psi_y^2 + \bar{f}_{TT\rho\rho}^{i+1,j}\psi_{1-x}^2 d\psi_y^2 - \bar{f}_{TT\rho\rho}^{i,j+1}\psi_x^2 d\psi_{1-y}^2 - \bar{f}_{TT\rho\rho}^{i+1,j+1}\psi_{1-x}^2 d\psi_{1-y}^2)/d\rho
\end{aligned}
$$

It might happen higher-order derivatives ($f_{TT\rho}$, $f_{T\rho\rho}$, $f_{TT\rho\rho}$) are not available in the LUT. In that case, sufficiently accurate finite-difference (FD) formulae can be used to approximate these HOD. A good strategy to preserve a compact stencil for the FD formulae while achieving high accuracy (*e.g.* 6th order on a 3×3 stencil) is to make use of available first and second derivatives in the LUT.

2.2 *Compact and Accurate Estimate of High-Order Derivatives*

Let us introduce the difference and arithmetic average FD operator in each direction of the LUT: $\delta_1 f_{i,j} = f_{i+\frac{1}{2},j} - f_{i-\frac{1}{2},j}$, $\mu_1 f_{i,j} = \frac{1}{2}(f_{i+\frac{1}{2},j} + f_{i-\frac{1}{2},j})$ for the first direction and similarly for the second direction. These difference operators can also be combined to yield the second-order accurate centered approximations of first and second-order derivatives:

$$
\frac{\delta_1 \mu_1 f_{i,j}}{\Delta T} = \frac{f_{i+1,j} - f_{i-1,j}}{2\Delta T} = (f_T)_{i,j} + \frac{\Delta T^2}{6}(f_{3T})_{i,j} + \frac{\Delta T^4}{120}(f_{5T})_{i,j} + O(\Delta T^6) \quad (2)
$$

Composing the difference operator with itself yields a second-order approximation of the second derivative in the i or T direction:

$$
\frac{\delta_1^2 f_{i,j}}{\Delta T^2} = (f_{TT})_{i,j} + \frac{\Delta T^2}{12}(f_{4T})_{i,j} + \frac{\Delta T^4}{360}(f_{6T})_{i,j} + O(\Delta T^6) \quad (3)
$$

and similarly in the j or ρ direction. The second-order FD centered formula for the mixed second-derivative reads:

$$\frac{\delta_1\mu_1\delta_2\mu_2 f_{i,j}}{\Delta T\Delta\rho} = (f_{T\rho})_{i,j} + \frac{\Delta T^2}{6}\left(f_{3T\rho}\right)_{i,j} + \frac{\Delta\rho^2}{6}\left(f_{T3\rho}\right)_{i,j} + \frac{\Delta T^4}{120}\left(f_{5T\rho}\right)_{i,j}$$
$$+\frac{\Delta T^2\Delta\rho^2}{36}\left(f_{3T3\rho}\right)_{i,j} + \frac{\Delta\rho^4}{120}\left(f_{T5\rho}\right)_{i,j} + O(h^6) \quad (4)$$

where $h = O(\Delta T, \Delta\rho)$. Let us assume the third and fourth-order derivatives appearing in the polynomial expansion (1) are not available in the LUT: our goal is to provide an accurate FD estimate for this derivative using the data available in the LUT. A straightforward way to proceed is to approximate for instance $f_{T2\rho}$ using the tabulated values of f and applying a FD formula for the first-derivative with respect to T combined with a FD formula for the second-derivative with respect to ρ. This first way to approximate $f_{T2\rho}$ reads $f^{(0)}_{T2\rho} = \frac{\delta_1\mu_1\delta_2^2 f}{\Delta T\Delta\rho^2}$ with the following corresponding Taylor development:

$$f^{(0)}_{T2\rho} = (f_{T2\rho}) + \frac{\Delta T^2}{6}(f_{3T2\rho}) + \frac{\Delta\rho^2}{12}(f_{T4\rho}) + \frac{\Delta T^4}{120}(f_{5T2\rho}) + \frac{\Delta T^2\Delta\rho^2}{72}(f_{3T4\rho})$$
$$+\frac{\Delta\rho^4}{360}(f_{T6\rho}) + O(h^6) \quad (5)$$

As observed from (5), using formula $f^{(0)}_{T2\rho}$ to estimate the mixed derivative $f_{T2\rho}$ ensures only a second-order accuracy. There are actually 4 other different ways to approximate $f_{T2\rho}$ using the tabulated values of the successive derivatives of f. It is indeed possible to compute $f_{T2\rho}$ as $(f_T)_{2\rho}$, that is the second-derivative with respect to ρ of f_T. It is also possible to compute $f_{T2\rho}$ as $(f_\rho)_{T\rho}$, that is the second mixed derivative of f_ρ; other choices are $(f_{2\rho})_T$, $(f_{T\rho})_\rho$. These various choices make respectively use of the tabulated values of f_T, f_ρ, $f_{2\rho}$ and $f_{T\rho}$. Corresponding Taylor developments read:

$$\begin{cases} f^{(1)}_{T2\rho} = \dfrac{\delta_2^2(f_T)}{\Delta\rho^2} = (f_{T2\rho}) + \dfrac{\Delta\rho^2}{12}(f_{T4\rho}) + \dfrac{\Delta\rho^4}{360}(f_{T6\rho}) + O(h^6) \\ f^{(2)}_{T2\rho} = \dfrac{\delta_1\mu_1\delta_2\mu_2(f_\rho)}{\Delta T\Delta\rho} = (f_{T2\rho}) + \dfrac{\Delta T^2}{6}(f_{3T2\rho}) + \dfrac{\Delta\rho^2}{6}(f_{T4\rho}) + \dfrac{\Delta T^4}{120}(f_{5T2\rho}) \\ \qquad + \dfrac{\Delta T^2\Delta\rho^2}{36}(f_{3T4\rho}) + \dfrac{\Delta\rho^4}{120}(f_{T6\rho}) + O(h^6) \\ f^{(3)}_{T2\rho} = \dfrac{\delta_1\mu_1(f_{2\rho})}{\Delta T} = (f_{T2\rho}) + \dfrac{\Delta T^2}{6}(f_{3T2\rho}) + \dfrac{\Delta T^4}{120}(f_{5T2\rho}) + O(\Delta h^6) \\ f^{(4)}_{T2\rho} = \dfrac{\delta_2\mu_2(f_{T\rho})}{\Delta\rho} = (f_{T2\rho}) + \dfrac{\Delta\rho^2}{6}(f_{T4\rho}) + \dfrac{\Delta\rho^4}{120}(f_{T6\rho}) + O(h^6) \end{cases} \quad (6)$$

The idea is now to look for an estimate of $f_{T2\rho}$ using a linear combination of the 5 previous formulae such that the second and fourth-order errors terms are cancelled out. After some straightforward algebra, we obtain:

$$f_{T2\rho} = -2f^{(0)}_{T2\rho} + 2f^{(1)}_{T2\rho} + f^{(2)}_{T2\rho} + f^{(3)}_{T2\rho} - f^{(4)}_{T2\rho} + O(h^6)$$

or else, detailing the corresponding FD formulae applied to f and its successive derivatives and which use the tabulated values in the LUT:

$$-2\frac{\delta_1\mu_1\delta_2^2(f)_{i,j}}{\Delta T\,\Delta\rho^2}+2\frac{\delta_2^2(f_T)_{i,j}}{\Delta\rho^2}+\frac{\delta_1\mu_1\delta_2\mu_2(f_\rho)_{i,j}}{\Delta T\,\Delta\rho}+\frac{\delta_1\mu_1(f_{2\rho})_{i,j}}{\Delta T}-\frac{\delta_2\mu_2(f_{T\rho})_{i,j}}{\Delta\rho}=(f_{T2\rho})_{i,j}+O(h^6) \tag{7}$$

The proposed 6th-order accurate discretization of $f_{T2\rho}$ on a 3×3 grid support introduces a strong improvement in accuracy with respect to the baseline formula (5). Similar ideas can be applied to derive a 3×3 sixth-order accurate FD discretization for the third-order mixed derivative $f_{2T\rho}$ as well as a 3×3 sixth-order accurate FD discretization for the fourth-order mixed derivative $f_{2T2\rho}$. Equipped with these high-accurate formulae, it is possible to apply the full bi-quintic Hermitian polynomial expansion (1) to interpolate thermodynamic quantities in the LUT. Figure 2 illustrates the accuracy and stability brought by the full bi-quintic interpolation including HOD with respect to the partial bi-quintic interpolation with HOD excluded when interpolating f_{TT} with f the Helmholtz free energy for the relativistic electron-positron gas EoS detailed in [1].

3 Preliminary Results

The FAST shocktube configuration described in [5] is reproduced using either the analytical Perfect Gas EoS or a LUT containing tabulated values of $f(T,\rho)$ and all its partial derivatives up to fourth-order so that the full bi-quintic Hermitian interpolation can be applied and compared with a conventional bilinear interpolation, which does not ensure thermodynamic consistency since it is applied separately for interpolating each physical quantity (pressure, speed of sound, etc). The quasi-1D Euler equations are discretized using the third-order Roe-MUSCL scheme. Figure 3 displays the convergence history of the 3 simulations (analytical EoS and LUT with either bi-quintic full interpolation or conventional bilinear interpolation) for a fine LUT containing 400×400 values of T and ρ. Since a dual-time step implicit

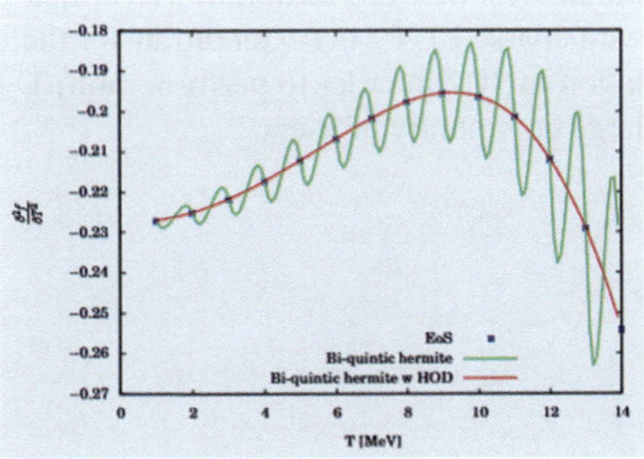

Fig. 2 Blue symbols: exact values of f_{TT} with f the Helmholtz free energy for the relativistic electron-positron gas EoS detailed in [1]. Red/green curves: full/partial bi-quintic interpolation including/excluding HOD

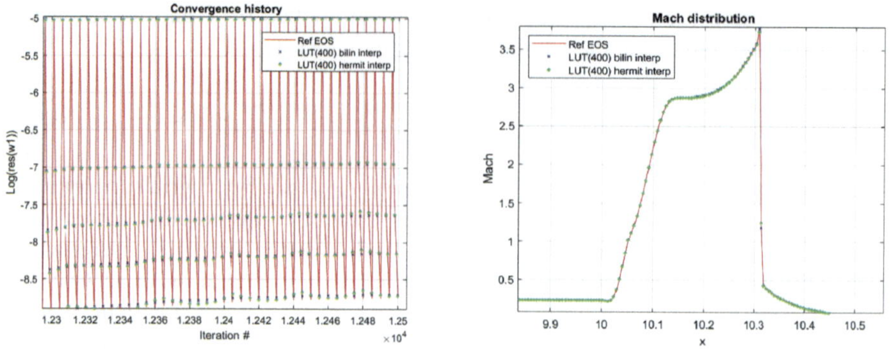

Fig. 3 Left: convergence history of the dual-time simulation of the FAST shock tube using the analytical Perfect Gas EoS or a fine 400 × 400 LUT for $f(T, \rho)$ combined either with conventional bilinear interpolation or a full bi-quintic Hermitian interpolation. Right: Mach number distribution in the FAST shock tube at $t = 2.5$ for the 3 numerical strategies

time-integration strategy is applied with 5 sub-iterations on dual time at each physical time-step (with $\Delta t = 10^{-3}$) and a large dual CFL number $CFL_\tau = 100$, one expects a fast convergence to the dual steady-state for each physical time-level. It can be checked such a fast convergence is indeed achieved using the analytical EoS but also with the LUT table whatever the interpolation strategy. Moreover, the Mach number distributions obtained at the final time $t = 2.5$ appear superimposed whatever the strategy followed to apply the EoS (analytical or LUT with Hermitian or bilinear interpolation). The benefit of using the Hermitian interpolation rather than the bilinear interpolation is evidenced in Fig. 4 where a close-up of the Mach distribution at $t = 2.5$ is displayed both for the previous fine 400 × 400 LUT and a coarse 100 × 100 LUT. While the bilinear interpolation is unable to preserve the accuracy of the simulation for such a coarse LUT, the consistent high-order Hermitian interpolation still provides a distribution superimposed with the distribution computed using the analytical EoS.

The next steps of the present work are: (i) perform a similar analysis with the cubic PRSV EOS before moving to more complex EOS; (ii) assess/confirm for these EOS the interest of a full bi-quintic Hermitian polynomial interpolation when combined with coarse LUT; (iii) extend/transfer the implementation of the Hermitian interpolation to SU2 in order to perform multi-D testcases and make the tool available to a large community of users.

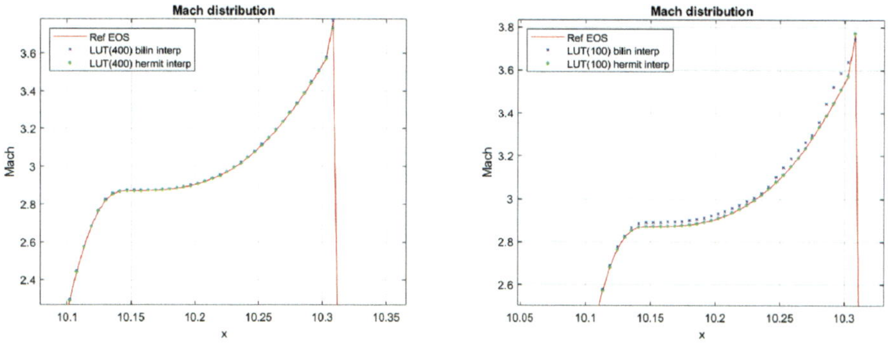

Fig. 4 Close-up of the final ($t = 2.5$) Mach number distribution in the FAST shock tube computed with the analytical Perfect Gas EoS or using a LUT for $f(T, \rho)$ either with bi-quintic interpolation or with bilinear interpolation. Left: fine 400×400 LUT. Right: coarse 100×100 LUT

References

1. Douglas Swesty, F.: Thermodynamically consistent interpolation for equation of state tables. J. Comput. Phys. **127**, 118–127 (1996)
2. Michal, Z., Milan, H., Pavel, V.: Hereos: A framework for consistent treatment of the equation of state in ale hydrodynamics. Comput. Math. Appl. **78**, 483–503 (2019)
3. Alberto, G., Piero, C., Matteo, P., Andrea, S.: Nonideal compressible fluid dynamics of dense vapors and supercritical fluids. Ann. Rev, Fluid Mech (2024)
4. Rubino, A., Pini, M., Kosec, M., Vitale, S., Colonna, P.: A look-up table method based on unstructured grids and its application to non-ideal compressible fluid dynamic simulations. J. Comput. Sci. **28**, 70–77 (2018)
5. Congedo, P.M., Colonna, P., Corre, C., Witteveen, J.A.S., Iaccarino, G.: Backward uncertainty propagation method in flow problems: Application to the prediction of rarefaction shock waves. Comput. Methods Appl. Mech. Eng. **314–326**(3), 213–216 (2012)

An Assessment of Navier-Stokes Characteristic Boundary Conditions for Real Fluids in a Non-Dissipative Numerical Framework

Zhen Li, Ahmed Abdellatif, Lluís Jofre, and Francesco Capuano

Abstract The abrupt variations in thermophysical properties pose significant challenges for the numerical simulation of real-fluid flows. This study evaluates Navier-Stokes characteristic boundary conditions (NSCBC) for non-ideal fluids in conjunction with kinctic-energy preserving (KEP) and pressure-equilibrium-preserving (PEP) schemes. The characteristic wave relations are applied to the total energy and pressure equations respectively. By demonstrating one-dimensional acoustic wave propagation, the profiles are shown to be significantly more stable for the real-gas regime in the combined KEP with PEP than that solely in KEP.

Keywords Real fluid · Characteristic wave analysis · Boundary condition · Kinetic-energy-preserving scheme · Pressure-equilibrium-preserving scheme

1 Introduction

High-pressure environments are prevalent in energy and propulsion fields, like for example in thermal-power cycles [1] and liquid rocket engines [2], as well as in advanced applications in microconfined turbulence [3, 4], leading working fluids to reach supercritical states and thus to present real-fluid effects. Due to the challenges associated with high-pressure conditions, there are arguably considerable limitations in the development of experimental techniques for non-ideal compressible

Z. Li
National University of Defense Technology, Changsha, China
e-mail: lizhen21@nudt.edu.cn

A. Abdellatif · L. Jofre · F. Capuano (✉)
Universitat Politècnica de Catalunya, Barcelona, Spain
e-mail: francesco.capuano@upc.edu

A. Abdellatif
e-mail: ahmed.abdellatif@upc.edu

L. Jofre
e-mail: lluis.jofre@upc.edu

A. Giauque et al. (eds.), *Proceedings of the 5th International Seminar on Non-Ideal Compressible Fluid Dynamics for Propulsion and Power*, ERCOFTAC Series 32,
https://doi.org/10.1007/978-3-031-92695-2_11

fluid dynamics [5]. Consequently, high-fidelity numerical simulations are nowadays considered an essential tool to gather insights into the physics of real-fluid flows, see, e.g., the recent review by Cinnella and Gloerfelt [6].

In the vicinity of the Widom line [7], the thermophysical properties of real fluids change dramatically, which requires a careful design of numerical schemes that are simultaneously non-dissipative and robust [8]. Furthermore, appropriate boundary conditions are also crucial to the accuracy and convergence of numerical results.

The approach based on characteristic wave decomposition is a popular choice to tackle boundary conditions in compressible flow, as it allows explicit control of the waves entering and leaving the domain. Initially proposed in the context of the Euler equations [9, 10], the characteristic boundary conditions were later applied to the Navier-Stokes equations by Poinsot and Lele [11] for perfect gases with constant and homogeneous thermodynamic properties. Since then, several extensions have followed. For instance, Baum [12] considered multi-component flows of thermally perfect gases. Okong'o and Bellan [13] were the first to extend Navier-Stokes characteristic boundary conditions (NSCBC) to real-gas flows using the local one-dimensional inviscid (LODI) assumption, in the context of an artificially-stabilized numerical framework.

Very recently, Williams [14] presented LODI-based NSCBC for hypersonic flows in thermochemical non-equilibrium.

Despite significant progress on NSCBC and novel numerical schemes, there is still need to obtain accurate conditions in conjunction with modern non-dissipative numerical methods with built-in stabilization for real fluids. For instance, recent works have utilized damping sponge layers to minimise disturbance reflections at the boundaries [15]. To this end, the objective of this work is to extend and assess characteristic-based boundary conditions for a recently-developed kinetic-energy- and pressure-equilibrium-preserving numerical framework [8]. The paper is organized as follows. In Sect. 2, the governing equations and discrete schemes for real fluids are introduced. Following that, the governing equations in the forms of characteristic waves are derived in Sect. 3. Then, tests using kinetic-energy-preserving (KEP) and pressure-equilibrium-preserving (PEP) schemes are reported in Sect. 4; concluding remarks are given in Sect. 5.

2 Methodology

2.1 Governing Equations of Real Fluids

The mass, momentum, and energy conservation equations for real fluids read

$$\frac{\partial \rho}{\partial t} + \frac{\partial (\rho u_i)}{\partial x_i} = 0\,, \tag{1}$$

$$\frac{\partial (\rho u_i)}{\partial t} + \frac{\partial (\rho u_i u_j)}{\partial x_j} + \frac{\partial p}{\partial x_i} = \frac{\partial \tau_{ij}}{\partial x_j}\,, \tag{2}$$

$$\frac{\partial (\rho E)}{\partial t} + \frac{\partial [(\rho E + p) u_i]}{\partial x_i} = -\frac{\partial q_i}{\partial x_i} + \frac{\partial (\tau_{ij} u_j)}{\partial x_i}\,, \tag{3}$$

where ρ is the density, u_i is the velocity along the x, y, and z directions ($i = 1, 2, 3$), p is the pressure, $\tau_{ij} = 2\mu A_{ij}$ is the viscous stress with μ the dynamic viscosity and $A_{ij} = \frac{1}{2}(\frac{\partial u_i}{\partial x_j} + \frac{\partial u_j}{\partial x_i}) - \frac{1}{3}\delta_{ij}\frac{\partial u_k}{\partial x_k}$, $E = e + \frac{1}{2}u_i u_i$ is the total energy (internal energy, e, plus kinetic energy), and q_i is the Fourier heat conduction flux.

The total energy equation can be substituted by an evolution equation for other quantities, leading to a system that (at the continuous level) is equivalent to the conservative one. In this work, the pressure evolution equation, Eq. (4) is utilized. It can be derived from Eq. (3), yielding

$$\frac{\partial p}{\partial t} + u_i \frac{\partial p}{\partial x_i} + \rho c^2 \frac{\partial u_i}{\partial x_i} = \frac{1}{\rho}\frac{\beta_v}{c_v \beta_T}\left(\tau_{ij}\frac{\partial u_j}{\partial x_i} - \frac{\partial q_i}{\partial x_i}\right), \tag{4}$$

where $c^2 = \left(\frac{\partial p}{\partial \rho}\right)_e + \frac{p}{\rho^2}\left(\frac{\partial p}{\partial e}\right)_\rho$ is the speed of sound, $\beta_v = \frac{1}{v}\left(\frac{\partial v}{\partial T}\right)_p$ is the volume expansivity, and $\beta_T = -\frac{1}{v}\left(\frac{\partial v}{\partial p}\right)_T$ is the isothermal compressibility.

2.2 *Equation of State*

The Peng-Robinson equation of state [16] is here employed to handle real-gas thermodynamics in the form

$$p = \frac{R_u T}{\bar{v} - b} - \frac{a}{\bar{v}^2 + 2b\bar{v} - b^2}, \tag{5}$$

where, R_u is the universal gas constant, $\bar{v} = W/\rho$ is the molar volume, and W is the molecular weight. Here, the coefficients a and b are defined as

$$a = 0.457\frac{(\mathrm{R_u} T_c)^2}{p_c}\left[1 + \tilde{c}\left(1 - \sqrt{T/T_c}\right)\right]^2 \quad \text{and} \quad b = 0.078\frac{\mathrm{R_u} T_c}{p_c}, \tag{6}$$

where coefficient $\tilde{c}$ is provided as a function of the acentric factor ω by

$$\tilde{c} = \begin{cases} 0.380 + 1.485\omega - 0.164\omega^2 + 0.017\omega^3 & \text{if} \quad \omega > 0.49 \\ 0.375 + 1.542\omega - 0.270\omega^2 & \text{otherwise.} \end{cases} \tag{7}$$

Although this work focuses on inviscid tests, the interested reader can refer to detailed work [17] for the calculation of the dynamic viscosity μ and thermal conductivity κ.

2.3 Discretization Framework

A standard semi-discretization procedure is employed for solving the equations of fluid motion as described in Bernades et al. [8], namely the equations are firstly discretized in space and then integrated in time. A second-order centered finite-differencing scheme is used for spatial discretization, while the fourth-order strong-stability preserving (SSP) Runge-Kutta approach is implemented in temporal integration. The convective terms are extended to ensure either KEP alone, or KEP and PEP simultaneously, as detailed in Ref. [8]. More in detail, the KEP scheme ensures that convective terms do not spuriously contribute to the discrete kinetic energy balance, by expanding them in the so-called KGP form; the PEP scheme also keeps the property of maintaining constant pressure when both pressure and velocity are initially uniform for real fluids, by evolving the pressure equation.

3 Characteristic Wave Analysis

3.1 Basics of LODI Assumption and NSCBC

In this work, the classical LODI assumption is employed, i.e., in developing the boundary conditions, the transverse and viscous terms are ignored and a one-dimensional system of governing equations is considered. When the conservative set of equations is employed, this reads

$$\frac{\partial \mathbf{U}}{\partial t} + \frac{\partial \mathbf{F}_1(\mathbf{U})}{\partial x_1} = \mathbf{0}, \tag{8}$$

where $\mathbf{U} = \left[\rho \ \rho u_1 \ \rho u_2 \ \rho u_3 \ \rho E\right]^{\mathbf{T}}$ and $\mathbf{F}_1(\mathbf{U})$ is the inviscid flux in the x direction.

Then, Eq. (8) is transformed into its primitive form

$$\frac{\partial \mathbf{V}}{\partial t} + \mathbf{A}^1 \frac{\partial \mathbf{V}}{\partial x_1} = \mathbf{0}, \tag{9}$$

Here, the matrix $\mathbf{A}^1$ can be diagonalized, $\mathbf{L}\mathbf{A}^1\mathbf{R} = \boldsymbol{\Lambda}^1$, where $\mathbf{L}$ and $\mathbf{R}$ are the left and right eigenvectors, $\boldsymbol{\Lambda}^1$ is the diagonal matrix. Defining $\mathcal{L} = \boldsymbol{\Lambda}^1\mathbf{L}\frac{\partial \mathbf{V}}{\partial x_1}$, then the Eq. (9) is written as Eq. (10) with $\mathbf{d} = \mathbf{R}\mathcal{L}$ as

$$\frac{\partial \mathbf{V}}{\partial t} + \mathbf{R}\mathcal{L} = \mathbf{0}. \tag{10}$$

The practical implementation of the procedure is summarized as follows:

- The solution $\mathbf{U}^n$ at time step n is known at the boundary; it is then possible to obtain the primitive variables $\mathbf{V}^n$.
- The right-hand side is calculated and modified at the boundary based on the *desired* wave amplitude. The amplitude variations of outgoing waves, $\mathcal{L} = \boldsymbol{\Lambda}^1\mathbf{L}\frac{\partial \mathbf{V}}{\partial x_1}$, are calculated from the interior points. The amplitude variations of incoming waves are calculated by Eq. (10).
- The solution at the boundary is updated at time $n+1$, $\mathbf{U}^{n+1}$, by means of Eq. (1).

3.2 Wave Amplitudes for Real Fluids

From the governing Eqs. (1), (2), and (4), it is possible to obtain the matrices $\mathbf{L}$, $\mathbf{R}$, and $\boldsymbol{\Lambda}$. Then

$$\mathcal{L} = \begin{pmatrix} (u_1 - c)\left(\frac{\partial p}{\partial x_1} - \rho c \frac{\partial u_1}{\partial x_1}\right) \\ u_1\left(c^2 \frac{\partial \rho}{\partial x_1} - \frac{\partial p}{\partial x_1}\right) \\ u_1 \frac{\partial u_2}{\partial x_1} \\ u_1 \frac{\partial u_3}{\partial x_1} \\ (u_1 + c)\left(\frac{\partial p}{\partial x_1} + \rho c \frac{\partial u_1}{\partial x_1}\right) \end{pmatrix}, \tag{11}$$

$$\mathbf{d} = \begin{pmatrix} \frac{1}{c^2}\left[\mathcal{L}_2 + \frac{1}{2}\left(\mathcal{L}_5 + \mathcal{L}_1\right)\right] \\ \frac{1}{2\rho c}\left(\mathcal{L}_5 - \mathcal{L}_1\right) \\ \mathcal{L}_3 \\ \mathcal{L}_4 \\ \frac{1}{2}\left(\mathcal{L}_5 + \mathcal{L}_1\right) \end{pmatrix}. \tag{12}$$

Therefore, Eqs. (1) and (4) can be written in the form of **d**, as follows

$$
\begin{aligned}
&\frac{\partial \rho}{\partial t}+d_1+\frac{\partial(\rho u_2)}{\partial x_2}+\frac{\partial(\rho u_3)}{\partial x_3}=0,\\
&\frac{\partial(\rho u_1)}{\partial t}+u_1 d_1+\rho d_2+\frac{\partial(\rho u_1 u_2)}{\partial x_2}+\frac{\partial(\rho u_1 u_3)}{\partial x_3}=\frac{\partial \tau_{1j}}{\partial x_j},\\
&\frac{\partial(\rho u_2)}{\partial t}+u_2 d_1+\rho d_3+\frac{\partial(\rho u_2 u_2)}{\partial x_2}+\frac{\partial(\rho u_2 u_3)}{\partial x_3}+\frac{\partial p}{\partial x_2}=\frac{\partial \tau_{2j}}{\partial x_j},\\
&\frac{\partial(\rho u_3)}{\partial t}+u_3 d_1+\rho d_4+\frac{\partial(\rho u_3 u_2)}{\partial x_2}+\frac{\partial(\rho u_3 u_3)}{\partial x_3}+\frac{\partial p}{\partial x_3}=\frac{\partial \tau_{3j}}{\partial x_j},\\
&\frac{\partial(\rho E)}{\partial t}+\left[\rho\left(\frac{\partial e}{\partial \rho}\right)_p+e+\frac{1}{2}u_k u_k\right]d_1+\rho u_1 d_2+\rho u_2 d_3+\rho u_3 d_4+\rho\left(\frac{\partial e}{\partial p}\right)_\rho d_5\\
&+\frac{\partial[(\rho E+p)u_2]}{\partial x_2}+\frac{\partial[(\rho E+p)u_3]}{\partial x_3}=-\frac{\partial q_i}{\partial x_i}+\frac{\partial(\tau_{ij}u_j)}{\partial x_i},
\end{aligned}
\tag{13}
$$

$$
\frac{\partial p}{\partial t}+d_5+u_2\frac{\partial p}{\partial x_2}+u_3\frac{\partial p}{\partial x_3}+\rho c^2\left(\frac{\partial u_2}{\partial x_2}+\frac{\partial u_3}{\partial x_3}\right)-\frac{1}{\rho}\frac{\beta_v}{c_v\beta_T}\left(\tau_{ij}\frac{\partial u_j}{\partial x_i}-\frac{\partial q_i}{\partial x_i}\right)=0.
\tag{14}
$$

For KEP, the total energy equation in Eq. 13 is utilized, while for KEP+PEP, the pressure Eq. 14 is employed.

3.3 *Boundary Conditions*

3.3.1 Subsonic Inflow

Imposing the velocity u_1, u_2, u_3, and temperature T, in the subsonic regime, $\lambda_1 = u_1 - c < 0$, $\lambda_2 = \lambda_3 = \lambda_4 = u_1 > 0$, and $\lambda_5 = u_1 + c > 0$. As a consequence, $\mathcal{L}_1$ is calculated by the interior points, i.e., Eq. (11). $\mathcal{L}_2$ to $\mathcal{L}_5$ are obtained from Eq. (10). Here,

$$
\begin{aligned}
\mathcal{L}_5 &= \mathcal{L}_1 - 2\rho c\frac{\partial u_1}{\partial t},\\
\mathcal{L}_3 &= -\frac{\partial u_2}{\partial t},\\
\mathcal{L}_4 &= -\frac{\partial u_3}{\partial t},\\
\mathcal{L}_2 &= \frac{1}{2}\left[\frac{c^2}{(\partial p/\partial \rho)_T}-1\right](\mathcal{L}_5+\mathcal{L}_1)+\frac{c^2}{(\partial p/\partial \rho)_T}\left(\frac{\partial p}{\partial T}\right)_\rho\frac{\partial T}{\partial t}.
\end{aligned}
\tag{15}
$$

3.3.2 Non-Reflecting Subsonic Outflow

For the subsonic outflow, $\mathcal{L}_2$ to $\mathcal{L}_5$ are calculated by the interior points. On the other hand, a relaxation condition is used for the incoming wave. As for $\mathcal{L}_1$

$$\mathcal{L}_1 = K\left(p - p_\infty\right), \tag{16}$$

where $K = \sigma\left(1 - M^2\right)c/L$; σ is a constant. When $\sigma = 0$, the boundary is perfectly non-reflecting. Here we set $\sigma = 0.3$ so that part of the physical information is enforced at the boundary.

4 One-Dimensional Acoustic Wave Propagation

We consider a one-dimensional acoustic wave propagation problem with nitrogen as the working fluid. The velocity, pressure, and density distributions correspond to

$$u = u_0 + A\exp\left[-\left(B\frac{x - L/2}{L}\right)^2\right], \tag{17}$$

$$p = p_0 + \rho_0 c_0\left(u - u_0\right), \tag{18}$$

$$\rho = \rho_0 + \frac{\rho_0(u - u_0)}{c_0}. \tag{19}$$

Here, the domain length $L = 1$ m, and is discretized with 201 equally-spaced grid points; the acoustic CFL is set to 0.3. The reference velocity u_0 is 1 m/s. The constants A and B, taking value of 5 and 10, determine the strength and stiffness of the acoustic wave, respectively.

4.1 Ideal-Gas Regime

In low-pressure, ideal-gas conditions, the reference pressure p_0 and reference density ρ_0 are set to 1 atm and 1.18 kg/m^3, respectively. The reference temperature T_0 is calculated by the ideal-gas equation of state. Figure 1 shows that the acoustic wave passes through the outflow boundary smoothly without noticeable disturbances whether the KEP scheme or the combined scheme of KEP and PEP. Upon more careful analysis, p-waves and packets of q-waves can be seen (not shown here), as previously reported in the literature [18, 19].

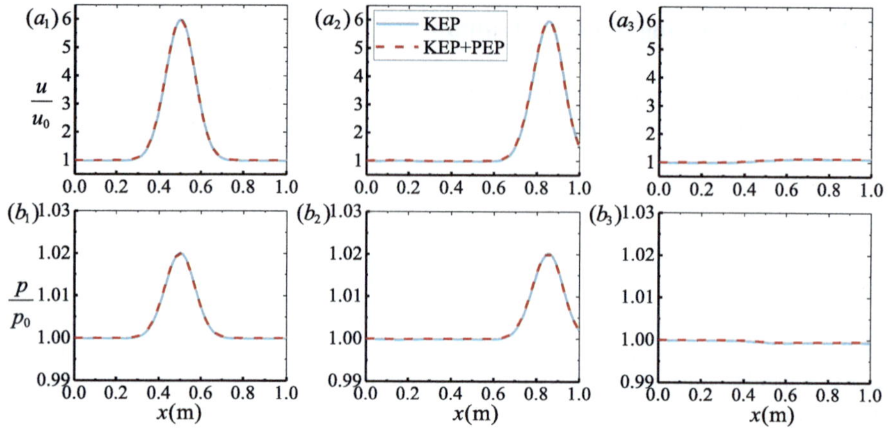

Fig. 1 Propagation of an acoustic wave in ideal-gas regime. **a** and **b** report the velocity and pressure respectively, which are reduced by the reference values. Subscripts 1, 2, and 3 represent $t_r = 0.0$, $t_r = 0.6934$, and $t_r = 2.0803$ with the reduced time $t_r = 2c_0t/L$

4.2 Real-Gas Regime

Results are also reported for real-gas conditions where the pressure is twice the critical pressure. At shown in Fig. 2, the results obtained from the KEP scheme display remarkable oscillations, whereas the combination of KEP and PEP yields stable results. This demonstrates the importance of pressure equilibrium for high-fidelity simulations of real fluids. The NSCBC conditions developed for the pressure

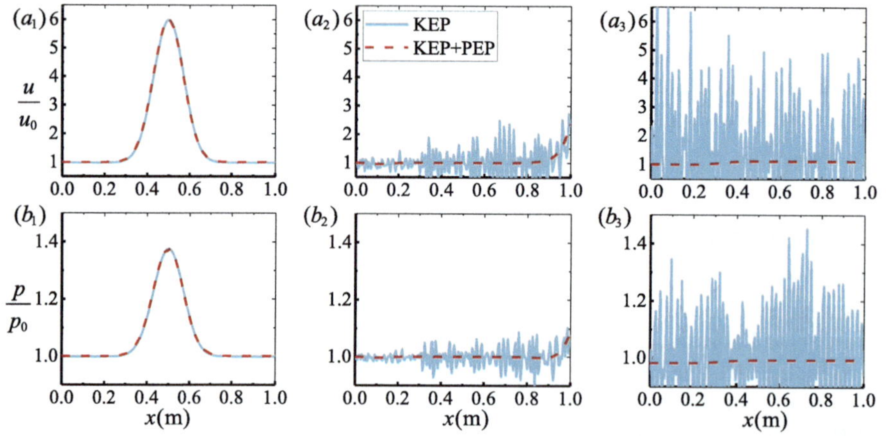

Fig. 2 Propagation of an acoustic wave in real-gas regime ($p = 2p_c$ and $T = 0.75T_c$). The notation of **a** and **b** is as shown in Fig. 1. Subscripts 1, 2, and 3 represent $t_r = 0.0$, $t_r = 1.2214$, and $t_r = 3.6643$

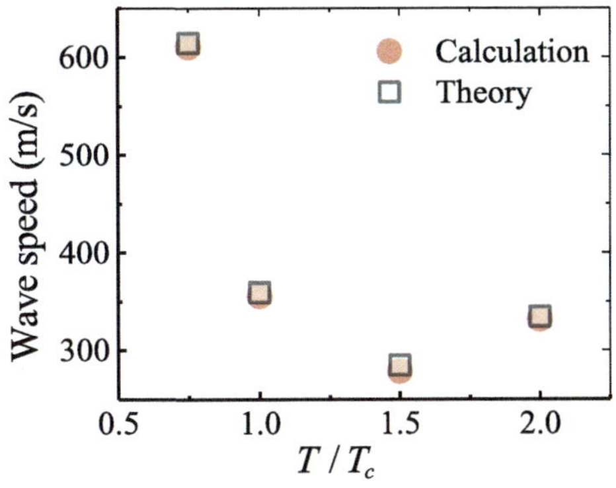

Fig. 3 Wave speed at different temperatures

equation also work very well. Finally, Fig. 3 presents the calculated wave speed at various temperature, which shows a good match with the theoretical results.

5 Conclusions and Future Work

This study has evaluated the effectiveness of NSCBC for real-fluid simulations, utilizing a recently developed KEP and PEP numerical framework. By applying characteristic wave relations to the pressure equation, the combined KEP and PEP approach provides more stable profiles for one-dimensional acoustic wave propagation in real fluid regimes compared to using KEP alone. This stability is crucial for accurately capturing the dynamics of high-pressure real fluids, thus enhancing the reliability and precision of numerical simulations in these contexts. Future research should focus on extending the evaluation of NSCBC with KEP and PEP schemes to more complex, multi-dimensional fluid flow scenarios.

Acknowledgements This research is supported by the China Scholarship Council (202306110026) and the Postgraduate Innovation Program of NUDT (XJJC2024058). The authors also acknowledge support from the *Serra Húnter* and SGR (2021-SGR-01045) programs of the Generalitat de Catalunya (Spain).

References

1. White, M.T., Bianchi, G., Chai, L., Tassou, S.A., Sayma, A.I.: Review of supercritical CO_2 technologies and systems for power generation. Appl. Therm. Eng. **185**, 116447 (2021)
2. Shanmugam, A.R., Park, K.S.: Flow and heat transfer of supercritical hydrogen in a regenerative cooling channel with the arc ribs of a rocket engine. Appl. Therm. Eng. **236**, 121451 (2024)
3. Bernades, M., Capuano, F., Jofre, L.: Microconfined high-pressure transcritical fluid turbulence. Phys. Fluids **35**(1) (2023)

4. Bernades, M., Jofre, L.: Thermophysical analysis of microconfined turbulent flow regimes at supercritical fluid conditions in heat transfer applications. J. Heat Transfer **144**(8), 082501 (2022)
5. Aus der Wiesche, S.: Experimental investigation techniques for non-ideal compressible fluid dynamics. Int. J. Turbomach. Propuls. Power **8**(2), 11 (2023)
6. Cinnella, P., Gloerfelt, X.: Insights into the turbulent flow of dense gases through high-fidelity simulations. Comput. Fluids 106067 (2023)
7. Banuti, D.T.: Crossing the Widom-line-supercritical pseudo-boiling. J. Supercrit. Fluids **98**, 12–16 (2015)
8. Bernades, M., Jofre, L., Capuano, F.: Kinetic-energy-and pressure-equilibrium-preserving schemes for real-gas turbulence in the transcritical regime. J. Comput. Phys. **493**, 112477 (2023)
9. Thompson, K.W.: Time dependent boundary conditions for hyperbolic systems. J. Comput. Phys. **68**(1), 1–24 (1987)
10. Thompson, K.W.: Time-dependent boundary conditions for hyperbolic systems, ii. J. Comput. Phys. **89**(2), 439–461 (1990)
11. Poinsot, T.J., Lele, S.K.: Boundary conditions for direct simulations of compressible viscous flows. J. Comput. Phys. **101**(1), 104–129 (1992)
12. Baum, M., Poinsot, T., Thévenin, D.: Accurate boundary conditions for multicomponent reactive flows. J. Comput. Phys. **116**(2), 247–261 (1995)
13. Okong'o, N., Bellan, J.: Consistent boundary conditions for multicomponent real gas mixtures based on characteristic waves. J. Comput. Phys. **176**(2), 330–344 (2002)
14. Williams, C., Di Renzo, M., Urzay, J., Moin, P.: Navier-Stokes characteristic boundary conditions for high-enthalpy compressible flows in thermochemical non-equilibrium. J. Comput. Phys. 113040 (2024)
15. Carlo Boldini, P., Gaspar, R., Pecnik, R., Bugeat, B.: Direct numerical simulation of H-type transition in a flat-plate boundary layer with supercritical fluids. In: The 14th International ERCOFTAC Symposium on Engineering Turbulence Modelling and Measurements (2023)
16. Peng, D.Y., Robinson, D.B.: A new two-constant equation of state. Ind. Eng. Chem. Fundam. **15**(1), 59–64 (1976)
17. Jofre, L., Urzay, J.: Transcritical diffuse-interface hydrodynamics of propellants in high-pressure combustors of chemical propulsion systems. Prog. Energy Combust. Sci. **82**, 100877 (2021)
18. Poinsot, T., Veynante, D.: Theoretical and numerical combustion. Edwards (2005)
19. Vichnevetsky, R., Bowles, J.B.: Fourier analysis of numerical approximations of hyperbolic equations. SIAM (1982)

Reacting Flows

Numerical Study of Droplet/flame Interaction Using a Diffuse Interface Method

Benoît Péden, Pierre Boivin, and Nicolas Odier

Abstract The interaction between a fuel droplet and a premixed laminar flame is investigated in this study using a multi-fluid Diffuse-Interface approach. The influence of two parameters on the reacting zone is examined: the equivalence ratio of the premixed laminar flame and the relative velocity between the droplet and the stabilized flame front. After validating the equation of state's extension to temperature-dependent heat capacities, direct numerical simulations of droplet-flame interaction are performed. The results are qualitatively analyzed, revealing a significant influence of the operating conditions.

Keywords Two-phase flows · Combustion · Evaporation · Diffuse interface method

1 Introduction

The combustion of sprays is characteristic of aeronautical gas turbines, leading to intricate spray/combustion interactions. Liquid injection and atomization processes are crucial as they determine the distribution of fuel vapor within a combustion chamber [1–3], significantly impacting the flame regime, burner efficiency, and pollutant emissions [4]. In such combustion chambers, liquid droplets of high inertia often cross the flame front [5]. In order to understand the interactions between an evaporating fuel droplet and a premixed flame, numerical simulations are mandatory.

Benoît Péden (✉)
CECI, Université de Toulouse, CERFACS/CNRS, Toulouse, France
e-mail: peden@cerfacs.fr

P. Boivin
Aix Marseille Univ, CNRS, Centrale Méditerrannée, M2P2, Marseille, France
e-mail: pierre.boivin@univ-amu.fr

N. Odier
CERFACS, Toulouse, France
e-mail: odier@cerfacs.fr

A. Giauque et al. (eds.), *Proceedings of the 5th International Seminar on Non-Ideal Compressible Fluid Dynamics for Propulsion and Power*, ERCOFTAC Series 32,
https://doi.org/10.1007/978-3-031-92695-2_12

Several approaches exist such as Direct Numerical Simulations (DNS) coupled with Discrete Particle Simulation (DPS) [1, 5, 6]. The latter method does not explicitly resolve the liquid phase. Several approaches have been proposed in the literature to conduct numerical simulation of fully eulerian two-phase flows. The method used in this study falls within this category and in particular in Multi-Fluid methods [7, 8].

The paper is structured as follows, (i) the methodology employed is described in Sect. 2 with the system of conservation equation, the thermodynamic closure and the phase-transition solver. (ii) The results are given in Sect. 3 where the method is validated on a canonical one dimensional flame and the study of droplet/flame interaction follows.

2 Methodology

This section presents the models for the simulation of compressible, reactive multicomponent flows in the AVBP solver [9, 10] from CERFACS. Initially, the system of conservation equations is introduced. Subsequently, the non-ideal thermodynamic closure and its extension for reactive flows are explained. Finally, the employed phase-transition solver is described.

2.1 *The 4-Equation Homogeneous Relaxation Method*

The 4-equation, or Homogeneous Relaxation Method, model [11] is employed to simulate compressible reactive and multicomponent flows. This model assumes mechanical and thermal equilibrium between phases in a computational cell, i.e., all components share the same pressure, velocity, and temperature. Consequently, a single set of conservation laws for mass, momentum, and energy is solved throughout the numerical domain, regardless of the fluid state (liquid or gas). This model, which belongs to the multi-fluid method [7, 8], treats the interfacial area as a numerically diffused zone. Under these assumptions, the 4-equation model considers a mono-component liquid ($k = 1$), its associated vapor ($k = 2$), and a multi-component, non-condensable gas mixture ($k = [3, \ldots, N]$), with N being the total number of species. The equilibrium conditions lead to the following variable definitions,

$$\begin{cases} T = T_k, \quad \forall k, \\ p = p_k, \quad \forall k, \\ v = 1/\rho = \sum_{k=1}^{N} Y_k v_k, \\ e = \sum_{k=1}^{N} Y_k e_k, \end{cases} \tag{1}$$

where Y_k represents the mass fraction of species k, T is the temperature, p is the pressure, ρ is the density, and e is the internal energy. The resulting four equations, with non-Euler right-hand side terms, are

$$\frac{\partial \rho}{\partial t} + \nabla \cdot (\rho u) = 0, \tag{2}$$

$$\frac{\partial \rho u}{\partial t} + \nabla \cdot (\rho u \otimes u + pI) = F_\sigma + \nabla \cdot (\mu \boldsymbol{\tau}), \tag{3}$$

$$\frac{\partial \rho E}{\partial t} + \nabla \cdot ((\rho E + p)u) = u \cdot F_\sigma + \nabla \cdot (\mu \boldsymbol{\tau} \cdot u) - \nabla \cdot q, \tag{4}$$

$$\frac{\partial \rho Y_k}{\partial t} + \nabla \cdot (\rho Y_k(u + v_k)) = \dot{\omega}_{c,k} + \dot{\omega}_{p,k}, \tag{5}$$

where u is the velocity vector and E is the total mixture energy ($E = e + u^2/2$). The right-hand side non-Euler term expressions are described in [12, 13]. The F_σ term represents the surface tension force, modelled using the Continuum Surface Force of Brackbill et al. [14], and reads $F_\sigma = \sigma \kappa \nabla z_\ell$. The σ parameter is the surface tension coefficient, κ is the curvature, computed using a color function [14], the liquid volume fraction here z_ℓ. The average mixture viscosity and heat conductivity are based on algebraic averages between the phases [12]. The source terms of Eq. 5 are denoted $\dot{\omega}_{c,k}$ for chemical reactions and $\dot{\omega}_{p,k}$ for phase transitions, detailed in Sect. 2.3. The diffusion effect for each component k is accounted for by introducing the diffusion velocity v_k.

2.2 The Extended Noble-Abel Stiffened-Gas Equation of State

Each component follows the Noble-Abel Stiffened-Gas (NASG) equation of state (EoS) [15]. This EoS is an improvement over the classic Stiffened-Gas EoS, accounting for repulsive molecular effects.

For a component k, the NASG EoS with temperature-variable heat capacities [16] is expressed as:

$$\begin{cases} v_k(p,T) = \dfrac{(C_{p,k}(T) - C_{v,k}(T))T}{p + p_{\infty,k}} + b_k, & (6a) \\ h_k(p,T) = \displaystyle\int C_{p,k}(T)\,dT + b_k p + q_k, & (6b) \\ e_k(p,T) = \displaystyle\int C_{v,k}(T)\,dT + \dfrac{(C_{p,k}(T) - C_{v,k}(T))T p_{\infty,k}}{p + p_{\infty,k}} + q_k, & (6c) \end{cases}$$

where $C_{p,k}$ and $C_{v,k}$ are the constant pressure and constant volume heat capacities, respectively; b_k is the covolume; and $p_{\infty,k}$ is a correction term to account for attraction (both terms are present only in the liquid phase, i.e., $k = 1$). The reference energy is denoted q_k. All these EoS parameters are determined by the thermodynamic properties of the fluid [16], not using the saturation curves [15, 17].

When using temperature-variable heat capacities, the mixture density ρ and internal energy e can be derived from the intensive variables p and T. However, the

Table 1 NASG EoS parameters employed in this study, with $\gamma = C_p/C_v$

Species	C_p [J/(kg.K)]	γ	q [J/kg]	b [m^3/kg]	p_∞ [Pa]
O_2	1005	1.349	−41 783	0	0
H_2O	2023	1.295	−78 766	0	0
N_2	1078	1.380	−18 777	0	0
KERO_LUCHE	2771	1.022	−754 009	0	0
CO	1090	1.374	−24 166	0	0
CO_2	1079	1.212	−139 174	0	0
$C_{10}H_{22,\ell}$	2386	1.216	-0.855 $\times 10^6$	9.4 $\times 10^{-4}$	2.875$\times 10^8$

$\int C_{v,k}(T)\,dT$ term in the energy equation of System (6) precludes explicit expressions for p and T from ρ and e. To address this issue, Boivin et al. [16] proposed a hybrid closure, stating that the constant heat capacity assumption only fails at high temperatures, where the liquid phase is no longer present. Heat capacities are assumed constant below a reference temperature ($T_{\text{ref}} = 700$ K), a temperature sufficiently high to ensure a mono-phase gaseous flow (far from the liquid saturation temperature). The hybrid closure posits that the EoS is NASG with constant heat capacities below T_{ref} and the classic perfect gas (PG) EoS above. The NASG EoS coefficients must be rederived to ensure $\mathscr{C}_1$ continuity at T_{ref}, in order to use the Janaf tables implemented in the AVBP solver. Originally, the NASG coefficients are derived from the NIST database [18], but the continuity of the enthalpy at T_{ref} is not guaranteed due to the choice of reference temperature, which is set to $T = 0$ K in the AVBP solver. The NASG parameters used in this paper are given in Table 1. The density ρ and the enthalpy h obtained with the NASG EoS coupled with the previously described methodology is plotted against NIST database curves [18] in Fig. 1, showing an accurate prediction in both liquid and gaseous states. The temperature $T_{\text{ref}} = 700$ K is thus justified as it is far from the saturation point (visible as a jump in thermodynamic quantities).

2.3 Phase Transition Model

The phase transition occurs between the liquid phase ($k = 1$) and its vapor phase ($k = 2$). The source terms apply exclusively to species 1 and 2 and are defined as:

$$\begin{cases} \dot{\omega}_{p,1} = \rho\nu(g_2 - g_1), \\ \dot{\omega}_{p,2} = -\rho\nu(g_2 - g_1), \end{cases} \tag{7}$$

where g_k denotes the Gibbs free energy of phase k, and ν is a function dependent on the specific interfacial area, pressure, and temperature. This parameter represents the relaxation rate of the mixture toward thermodynamic equilibrium. Within this frame-

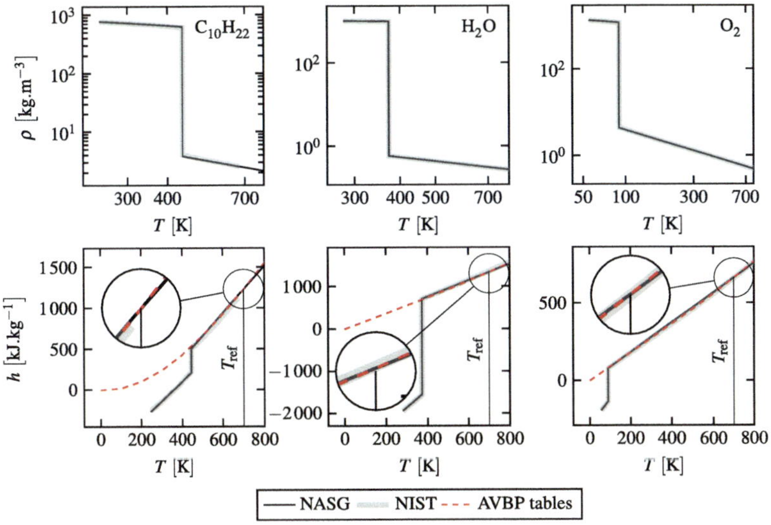

Fig. 1 Validation of the NASG parameters against the NIST database [18]. Profiles of density and enthalpy against temperature at $p = 1$ bar

work, an instantaneous relaxation of Gibbs free energies is assumed [19]. Finite-rate models are also documented in the literature [20, 21]. Instead of directly solving Eq. (7), the fast phase transition solver proposed by Chiapolino et al. [19] is utilized.

3 Results

3.1 Validation of the Implemented Hybrid Thermodynamic Closure Kerosene Surrogate Laminar Pre-Mixed Flame

A one-dimensional, freely propagating pre-mixed flame is examined using fuel closely related to the targeted aeronautical engines applications. The chemical mechanism employed is the well-known two-step kerosene BFER model [22], which includes 6 + 1 species, with the liquid $C_{10}H_{22}$ added. The initial conditions are specified in Table 2, at atmospheric pressure. A small amount of liquid kerosene $C_{10}H_{22,\ell}$ is uniformly distributed throughout the domain. The species KERO_LUCHE represents the vapor phase associated with the liquid $C_{10}H_{22}$.

The resulting solution is depicted in Fig. 2 and is compared with results obtained from the AVBP solver using the perfect gas EoS. The two sets of simulation results are superimposed, and no jump on the temperature/heat release curve appears at T_{ref}, thereby validating the implementation.

Table 2 Initial conditions for the one-dimensional free-propagating $C_{10}H_{22}$-Air pre-mixed flame.

Variable	T [K]	u [m s^{-1}]	$Y_{C_{10}H_{22},\ell}$	Y_{KERO_LUCHE}	Y_{CO_2}	Y_{CO}	Y_{O_2}	Y_{H_2O}	Y_{N_2}
Fresh gases	298	0.319	1×10^{-8}	5.775×10^{-2}	0	0	2.20×10^{-1}	0	$1 - \sum_{k=1}^{N-1} Y_k$
Burnt gases	2217	2.5	1×10^{-8}	0	1.733×10^{-1}	5.015×10^{-3}	2.482×10^{-2}	7.416×10^{-2}	$1 - \sum_{k=1}^{N-1} Y_k$

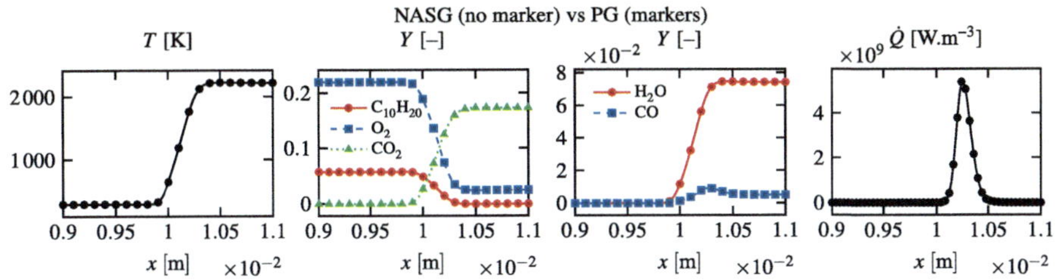

Fig. 2 Free-propagating kerosene flame. Temperature, species and heat release profiles for both equations of state: NASG without and PG with markers

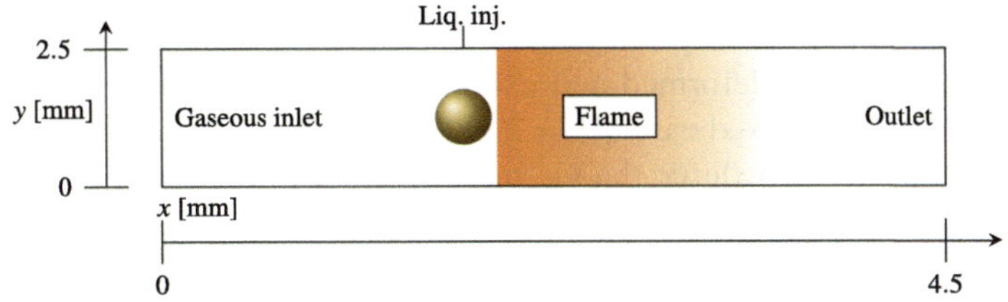

Fig. 3 Schematic of the configuration

3.2 Droplet/Laminar Flame Front Interaction

In this section, the interaction of an evaporating fuel droplet with a laminar premixed flame is investigated. The numerical framework, validated in the previous subsection is employed. The results are analyzed qualitatively. The laminar premixed flame equivalence ratio, as well as the relative velocity between the fuel droplet and the flame front, ($u^* = u_d/S_l$ where u_d is the droplet initial velocity and S_l is the laminar flame speed), are the two parameters of this study.

Configuration The computational domain, as depicted in Fig. 3, comprises a [4.5 mm × 2.5 mm] two-dimensional region constructed with 537×10^3 triangular elements. The boundary conditions, as detailed in [23], consist of a non-reflecting inlet on the left, a non-reflecting outlet on the right, both extended to NASG EoS [13], and two symmetries on the top and bottom boundaries. The chemistry is described using the 2-step mechanism, validated above, and reported in [22], encompassing $6 + 1$ species and 2 reactions. A third-order Galerkin scheme in space and time, TTGC [24], is applied. Additionally, the Localized Artificial Diffusion (LAD) method, as described in [25], is employed. The surface tension coefficient for n-decane is set at $\sigma = 24 \times 10^{-3}$ N/m. The gaseous mixture is injected at the flame front velocity. A liquid n-decane droplet of diameter $d = 80\,\mu$m is placed in front of the reacting zone, after the establishment of the purely gaseous premixed flame. The composition of the gaseous premixed flame, for each equivalence ratio, is determined considering the thermodynamic equilibrium condition with Cantera [26] for this chemical mechanism.

3.2.1 Influence of the Premixed Flame Equivalence Ratio

A droplet with a velocity equal to 10 times the premixed flame velocity is considered for two equivalence ratios, $\phi = 0.7$ and $\phi = 1.3$. For each initial equivalence ratio, the temperature and heat release solutions are presented in Fig. 4.

The behavior differs between the two cases: for $\phi = 0.7$, a non-zero mass fraction of O_2 (Y_{O_2}) is observed in the burnt gases zone. Consequently, the interaction of the evaporating fuel with this zone generates a second flame front, characterized by a lower heat release, located at the $\phi = 1$ iso-contour (Fig. 4 bottom left). This topology is reminiscent of diffusion flames. Conversely, for $\phi = 1.3$, no oxidizer is present in the burnt gases zone, causing the reaction to cease. It is also noteworthy that the passage of the evaporating droplet significantly affects the flame front for the $\phi = 0.7$ case; the flame front is deformed and exhibits a higher heat release compared to the purely gaseous premixed case ($\dot{Q} = 2 \times 10^9 \mathrm{W/m^3.s}$). The reaction in the burnt gases is evident in Fig. 5 (left): the O_2 mass fraction in the burnt gases, between the $\phi = 1$ iso-contour and the droplet, is zero. Additionally, the Takeno flame index (blue → non-premixed) evidences this diffusion mode in the burnt gases. Finally, the iso-contours of liquid volume fraction $z_\ell = 0.3$ (in red) in Fig. 5 exhibit a higher evaporation rate for the $\phi = 0.7$ case as the contour zone is smaller.

Influence of the relative velocity of the droplet to the flame front For the case $\phi = 0.7$, a droplet with the same velocity as the flame front ($u^* = 1$) is placed in front of the reacting zone. The global topologies of the temperature and the heat release field are retrieved, the temperature and heat release solutions are presented in Fig. 6. The same observation regarding a second heat release in the burnt gases

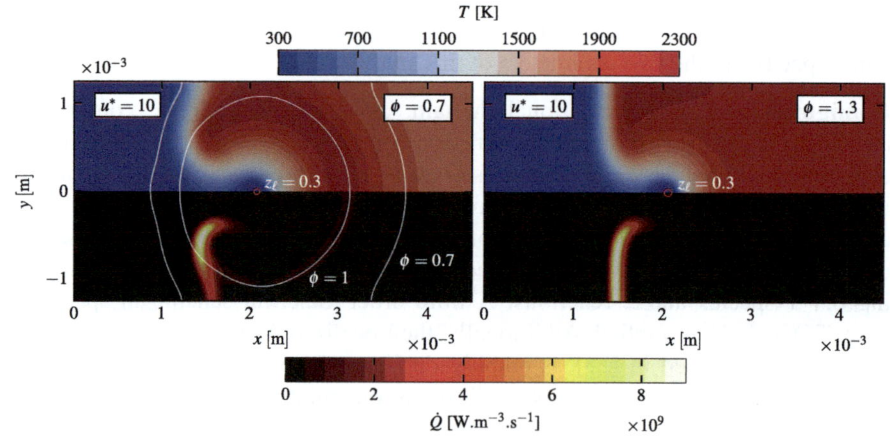

Fig. 4 Influence of the equivalence ratio of the premixed flame. Left: $\phi = 0.7$, right: $\phi = 1.3$. The top half of each figure represents the temperature field and the bottom half represents the heat release field. For the case initially at $\phi = 0.7$, iso-contours of $\phi = 0.7$ and $\phi = 1$ are shown. The red iso-contour represents the liquid volume fraction ($z_\ell = 0.3$)

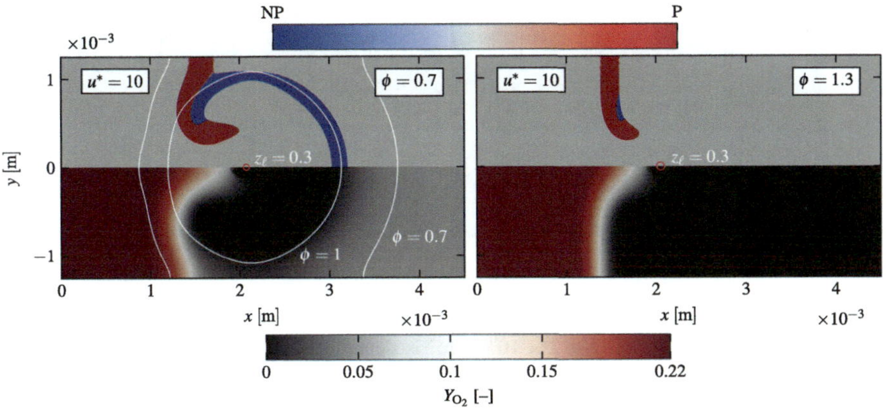

Fig. 5 Influence of the equivalence ratio of the premixed flame. Left: $\phi = 0.7$, right: $\phi = 1.3$. The top half of each figure represents the vapor fuel mass fraction and the bottom half represents the oxidizer mass fraction. For the case initially at $\phi = 0.7$, iso-contours of $\phi = 0.7$ and $\phi = 1$ are shown. The red iso-contour represents the liquid volume fraction ($z_\ell = 0.3$)

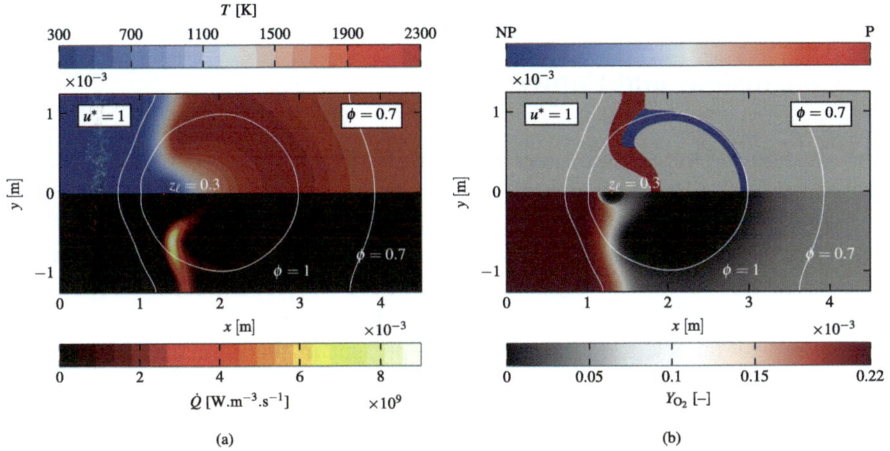

Fig. 6 Influence of the relative velocity of the droplet to the premixed flame for $\phi = 0.7$. **a** The top half represents the temperature field and the bottom half represents the heat release field. **b** The top half represents the Takeno flame index and the bottom half represents the O_2 species mass fraction field. Iso-contours of $\phi = 0.7$ and $\phi = 1$ are shown. The red iso-contour represents the liquid volume fraction ($z_\ell = 0.3$)

zone can be made: a high temperature zone is exhibited on the $\phi = 1$ iso-contour. However, as pointed out by the liquid volume fraction iso-contour ($z_\ell = 0.3$), the droplet does not cross the flame front due to its initial velocity.

Two instantaneous fields of liquid volume fraction (z_ℓ) are shown in Fig. 7. An iso-contour of $\dot{Q} = 2 \times 10^9 \mathrm{W/m^3}$ is also shown, which corresponds to the initial premixed flame heat release. It means that for the case at $u^* = 1$, the droplet is almost

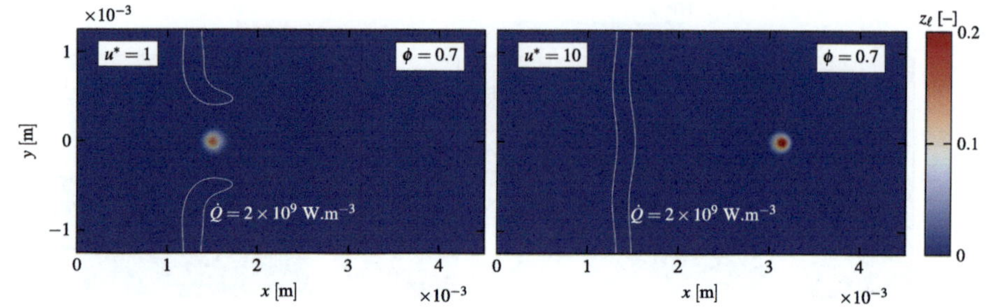

Fig. 7 Influence of the relative velocity of the droplet to the premixed flame for $\phi = 0.7$. Liquid volume fraction z_ℓ fields with $\dot{Q} = 2 \times 10^9 \mathrm{W/m^3}$ iso-contours (white lines)

completely vaporized before crossing the flame front. However, for the $u^* = 10$ case, the droplet crosses the flame front before being completely vaporized. For the latter case, the flame front retrieves the initial premixed value while the droplet finishes vaporizing in the burnt gases.

4 Conclusions and Future Works

The extension of the NASG equation of state (EoS) [15] to temperature-dependent heat capacities [16] has been implemented and validated within the AVBP solver. The interactions between droplets and flames have demonstrated significantly different behaviors when varying the following parameters: the premixed flame equivalence ratio ϕ and the relative velocity u^* between the fuel droplet and the initial premixed flame front. Physical insights were obtained: when the premixed flame equivalence ratio ϕ is less than 1, a diffusion flame is present in the burnt gases zone. This phenomenon is naturally absent when ϕ is greater than 1. For droplets with an initial velocity greater than the flame front, the droplet traverses the reacting zone and continues to evaporate in the burnt gases zone; the reacting zone then relaxes towards the initial purely gaseous premixed flame configuration. Conversely, when the droplet's velocity matches that of the flame front, the droplet nearly completely vaporizes before crossing the reaction zone. These encouraging results indicate that the method captures physical phenomena accurately and can be applied to future, more complex cases and more detailed DNS studies.

Acknowledgements This work was supported by the French National Research Agency through the project MADIRAN (grant ANR-21-CE50-0043) and benefited from GENCI-CINES HPC resources (grant 2023 A0152B10157).

References

1. Réveillon, J., Vervisch, L.: Analysis of weakly turbulent dilute-spray flames and spray combustion regimes. J. Fluid Mech. **537**(1), 317–347 (2005)
2. Luo, K., Pitsch, H., Pai, M.G., Desjardins, O.: Direct numerical simulations and analysis of three-dimensional n-heptane spray flames in a model swirl combustor. Proc. Combust. Inst. **33**(2), 2143–2152 (2011)
3. Sánchez, A.L., Urzay, J., Liñán, A.: The role of separation of scales in the description of spray combustion. Proc. Combust. Inst. **35**(2), 1549–1577 (2015)
4. Beck, C.H., Koch, R., Bauer, H.-J.: Identification of droplet burning modes in lean, partially prevaporized swirl-stabilized spray flames. Proc. Combust. Inst. **32**(2), 2195–2203 (2009)
5. F. Shum-Kivan, J. Marrero Santiago, A. Verdier, E. Riber, B. Renou, G. Cabot, and B. Cuenot. Experimental and numerical analysis of a turbulent spray flame structure. *Proceedings of the Combustion Institute*, 36(2):2567–2575, 2017
6. Rochette, B., Riber, E., Cuenot, B.: Effect of non-zero relative velocity on the flame speed of two-phase laminar flames. Proc. Combust. Inst. **37**(3), 3393–3400 (2019)
7. Saurel, R., Abgrall, R.: A multiphase godunov method for compressible multifluid and multi phase flows. J. Comput. Phys. **150**(2), 425–467 (1999)
8. Kapila, A., Menikoff, R., Bdzil, J., Son, S., Stewart, D.: Two-phase modeling of deflagration-to-detonation transition in granular materials: reduced equations. Phys. Fluids **13**(10), 3002–3024 (2001)
9. Schönfeld, T., Rudgyard, M.: Steady and unsteady flow simulations using the hybrid flow solver AVBP. AIAA J. **37**(11), 1378–1385 (1999)
10. Gicquel, L.Y.M., Gourdain, N., Boussuge, J.F., Deniau, H., Staffelbach, G., Wolf, P., Poinsot, T.: High performance parallel computing of flows in complex geometries. Comptes Rendus Mécanique **339**(2-3), 104–124 (2011)
11. Saurel, R., Boivin, P., Le Métayer, O.: A general formulation for cavitating, boiling and evaporating flows. Comput. Fluids **128**, 53–64 (2016)
12. Deng, X., Boivin, P.: Diffuse interface modelling of reactive multi-phase flows applied to a sub-critical cryogenic jet. Appl. Math. Model. **84**, 405–424 (2020)
13. Péden, B., Carmona, J., Boivin, P., Schmitt, T., Cuenot, B., Odier, N.: Numerical assessment of diffuse-interface method for air-assisted liquid sheet simulation. Comput. Fluids **266**, 106022 (2023)
14. Brackbill, J.U., Kothe, D.B., Zemach, C.: A continuum method for modeling surface tension. J. Comput. Phys. **100**(2), 335–354 (1992)
15. Le Métayer, O., Saurel, R.: The noble–abel stiffened–gas equation of state. Phys. Fluids **28**(4), 046102 (2016)
16. Boivin, P., Cannac, M.A., Le Métayer, O.: A thermodynamic closure for the simulation of multiphase reactive flows. Int. J. Thermal Sci. **137**, 640–649 (2019)
17. Le Métayer, O., Massoni, J., Saurel, R.: Elaboration des lois d'état d'un liquide et de sa vapeur pour les modéles d'écoulements diphasiques. Int. J. Thermal Sci. **43**(3), 265–276 (2004)
18. Linstrom, P.J., Mallard, W.G.: The NIST chemistry WebBook: a chemical data resource on the internet. J. Chem. Eng. Data **46**(5), 1059–1063 (2001)
19. Chiapolino, A., Boivin, P., Saurel, R.: A simple and fast phase transition relaxation solver for compressible multicomponent two-phase flows. Comput. Fluids **150**, 31–45 (2017)
20. Furfaro, D., Saurel, R.: Modeling droplet phase change in the presence of a multi-component gas mixture. Appl. Math. Comput. **272**, 518–541 (2016)
21. Pelanti, M.: Arbitrary-rate relaxation techniques for the numerical modeling of compressible two-phase flows with heat and mass transfer. Int. J. Multiphase Flow **153**, 104097 (2022)
22. Franzelli, B., Riber, E., Sanjosé, M., Poinsot, T.: A two-step chemical scheme for keroseneair premixed flames. Combust. Flame **157**(7), 1364–1373 (2010)
23. Poinsot, T.J., Lele, S.K.: Boundary conditions for direct simulations of compressible viscous flows. J. Comput. Phys. **101**(1), 104–129 (1992)

24. Colin, O., Rudgyard, M.: Development of high-order taylor–galerkin schemes for LES. J. Comput. Phys. **162**(2), 338–371 (2000)
25. Schmitt, T.: Large-eddy simulations of the mascotte test cases operating at supercritical pressure. Flow Turbul. Combust. **105**, 159–189 (2020)
26. Goodwin, D.G., Moffat, H.K., Speth, R.L.: Cantera - An object-oriented software toolkit for chemical kinetics, thermodynamics, and transport processes. (2017). http://www.cantera.org

Evaluation of a Multi-fluid Method for Reactive Subcritical Rocket Engine Using Large Eddy Simulation

Thibault Gioud, Nicolas Odier, and Thomas Schmitt

Abstract Rocket engines may face various thermodynamic conditions, including subcritical regimes where liquid and gaseous phases coexist in a reactive, compressible environment. In this context, a multi-fluid approach, which assumes equilibrium of temperature, pressure, velocity, and Gibbs potentials, is considered to simulate these flows. Surface tension forces, which play a crucial role in atomization phenomena, are accounted for in this work. The method is evaluated on a reactive configuration representative of rocket engine conditions (test bench investigated at Technische Universität München (TUM)), presenting very promising results and highlighting the benefits of simulating the liquid jet compared to a method that models it through the injection of Lagrangian particles.

Keywords Two-phase flows · Multi-fluid · Rocket engine · LES

1 Introduction

In recent years, the propellant combination of liquid oxygen (LOx) and methane (CH4) has received a lot of interest for future rocket applications. In this context, the *Technische Universität München* (TUM) has developed various sub-scale rocket combustors that use methane as fuel. Numerous firing tests have been conducted using the GOx/GCH4 combination, exploring various configurations such as different section shapes and the number of injectors, as detailed in [1–3]. Additionally, other experimental setups involve a subcritical regime with liquid oxygen (LOx), as

T. Gioud (✉) · N. Odier
CERFACS, Toulouse, France
e-mail: gioud@cerfacs.fr

N. Odier
e-mail: odier@cerfacs.fr

T. Schmitt
EM2C, CNRS, CentraleSupélec, Université Paris-Saclay, Paris, France
e-mail: thomas.schmitt@centralesupelec.fr

A. Giauque et al. (eds.), *Proceedings of the 5th International Seminar on Non-Ideal Compressible Fluid Dynamics for Propulsion and Power*, ERCOFTAC Series 32,
https://doi.org/10.1007/978-3-031-92695-2_13

discussed in [4–9]. Our study focuses on a subcritical configuration (LOx/GCH4), introduced in [5].

Simulating two-phase flows presents significant numerical challenges due to the sharp gradients at the interface. Advanced numerical methods for two-phase flow have been developed to capture these interfaces, using accurate schemes that handle complex geometries. These methods are sharp interface approaches, such as the Volume of Fluid (VoF) method (Hirt and Nichols [10]), the Level-Set method (Sussman et al. [11]), and hybrid techniques like the Coupled VoF/Level-Set approach (Sussman and Puckett [12]). These methods, typically based on incompressible flow equations, have enabled the Direct Numerical Simulations (DNS) of jet atomization ([13–15]), yielding important physical insights, but generally at a significant computational cost. In the context of industrial reactive configurations, the two-phase flow model considered here is the Homogeneous Multi-fluid Method (HMM), which is considered as a diffuse interface approach.

The objective of this paper is to evaluate the stability and accuracy of the HMM, under realistic conditions of a liquid rocket engine. The multi-fluid method considered in this study has been previously assessed under subcritical conditions by [16, 17], but without accounting for surface tension. Furthermore, wall heat losses were not studied in their simulations. A numerical simulation employing a Lagrangian formalism to describe the liquid phase for the same configuration was conducted by [18]. This study yielded encouraging results, despite some limitations in predictability near the injection point and uncertainties associated with the Lagrangian injection model. The findings from this previous study will be used for comparison with the present results, alongside experimental data.

This paper is structured as follows: first, a description of the experimental setup and the numerical models being used is provided. Following this, results are shown, including a comparison with the Blanchard [18] simulation, run with a Lagrangian formalism, as well as an experimental comparison and a qualitative analysis of the phenomena.

2 Experimental Configuration

The configuration under study consists of a single LOx/GCH4 injector that was experimentally investigated under firing conditions at TUM [5]. Gaseous methane and liquid oxygen are injected into the combustion chamber through a single coaxial injector. The combustor is 290 mm long and has a square cross-section measuring 12x12 mm. The geometrical properties of the test rig are summarized in Table 2.

The operating conditions considered for the simulations are presented in Table 1. Oxygen is injected at a reduced temperature (T_r) and reduced pressure (P_r) below 1, indicating that the oxygen is in a subcritical and liquid state. Conversely, methane is injected with a reduced temperature (T_r) greater than 1 and a subcritical pressure, indicating that the methane is in a gaseous state.

Table 1 Operating conditions

Property	Value	Unit
Pressure at inlet	18.5	bar
LOx $\dot{m}$	44.63	g/s
GCH4 $\dot{m}$	14.9	g/s
Global equivalence ratio ϕ	1.33	–
LOx reduced pressure (Pr)	0.37	–
LOx reduced Temperature (Tr)	0.67	-
GCH4 reduced Temperature	1.42	–
GCH4 reduced pressure	0.4	–
$We_{liq} = \frac{\rho_l u_l D_{inj}}{\sigma}$	2200	–
$Re_{liq} = \frac{\rho_l u_l D_{inj}}{\mu}$	157 000	–
$Re_{gas} = \frac{\rho_g u_g D_{inj}}{\mu}$	530 000	–

Table 2 Geometrical parameters

Property	Value	Unit
Chamber length	290	mm
Chamber width	12	mm
Chamber height	12	mm
LOx injection diameter	3	mm
Lip thickness	1	mm
GCH4 injection inner diameter	5	mm
GCH4 injection outer diameter	6	mm

The experimental data acquisition was performed using 12 thermocouples and 9 pressure probes located along the axis passing through the middle of the upper face. Thermocouples are separated by a distance of 17 mm and the pressure probes by a distance of 34 mm. These probes allow to provide axial pressure and wall heat loss profiles. It is noteworthy that the heat fluxes measurement uncertainties lie around 8%.

3 Numerical Modelling

The numerical simulations are performed with the massively parallel, explicit Large-Eddy Simulation (LES) solver AVBP developed at CERFACS [19, 20]. The numerical scheme is a Two-step Taylor Galerkin method [21], 3rd order accurate in space and time. Unresolved scales are modelled thanks to the WALE Sub-Grid Scale (SGS) model [22]. Constant turbulent Schmidt and Prandtl numbers (fixed at 0.6 for both) are used for the sub-grid species and heat diffusion terms.

Assuming the equilibrium of pressure, velocity, temperature and chemical potential between the liquid and the gas, the 3-equation [16] homogeneous equilibrium for the Favre-filtered fully compressible Navier-Stokes equations yields to Eq. (1):

$$\frac{\partial \overline{\rho}\widetilde{Y}_k}{\partial t} + \nabla.(\bar{\rho}\widetilde{Y}_k\widetilde{\mathbf{u}}) = -\nabla\overline{\mathscr{J}}_k - \nabla\mathscr{J}_k^t + \bar{\dot{\omega}}_k \quad \text{with} \quad k = [1, ..., N_s] \tag{1a}$$

$$\frac{\partial \overline{\rho}\widetilde{\mathbf{u}}}{\partial t} + \nabla.(\overline{\rho}\widetilde{\mathbf{u}} \otimes \widetilde{\mathbf{u}} + \overline{P}\underline{\underline{\mathbf{I}}}) = \nabla(\mu\underline{\underline{\overline{\tau}}}) + \nabla\left(\mu\underline{\underline{\tau}}^t\right) \tag{1b}$$

$$\frac{\partial \overline{\rho}\widetilde{e}_{tot}}{\partial t} + \nabla.(\overline{\rho}e_{tot} + \overline{P})\widetilde{\mathbf{u}}) = \nabla(\mu\underline{\underline{\overline{\tau}}}.\widetilde{\mathbf{u}}) - \nabla\overline{\mathbf{q}} - \nabla q^t + \bar{\dot{\omega}}_T \tag{1c}$$

where $\bar{\cdot}$ and $\tilde{\cdot}$ are the spatial and mass-weighted Favre spatial filters, respectively. The density, velocity and pressure are respectively represented by ρ, $\mathbf{u}$ and P. The overall (gas + liquid) species mass fractions are $Y_k = m_k/m = \{Y_1, \ldots, Y_{N_s}\}$, N_s being the number of species. The viscous stress tensor is noted $\underline{\underline{\tau}}$ and the total energy is $e_{tot} = e_c + e_s$, with $e_c = \mathbf{u}^2/2$ the kinetic energy and e_s the sensible energy. The heat fluxes are denoted q. The subgrid scale fluxes, represented by the subgrid scale (SGS) stress tensor $\underline{\underline{\tau}}^t$, SGS energy fluxes q^t and SGS species fluxes $\mathscr{J}_k^t$, are considered in the formulation. Furthermore, the species reaction rate and heat release rate are denoted as $\dot{\omega}_k$ and $\dot{\omega}_T$, respectively. In this approach, the liquid phase and the gas phase are likely to coexist, and the mixture density and energy are defined in Eq. (2):

$$\rho = \alpha_l\rho_l^{SAT} + (1 - \alpha_l)\rho_v^{SAT} \tag{2a}$$

$$\rho e_s = \alpha_l\rho_l^{SAT}e_{s,l}^{SAT} + (1 - \alpha_l)\rho_v^{SAT}e_{s,v}^{SAT} \tag{2b}$$

where α_l is the liquid volume fraction defined in Eq. (3) and the superscript *SAT* represents the properties at saturation. Finally, subscripts l and v respectively denote the liquid and the vapour phase.

$$\alpha_l = \frac{\rho - \rho_g^{SAT}}{\rho_l^{SAT} - \rho_g^{SAT}} \tag{3}$$

In this work, the thermodynamic closure is based on the cubic equation of state introduced by Peng and Robinson [23] (PR), as described in equation Eq. (4):

$$P(\rho, T, \underline{Y}) = \frac{\rho\bar{r}T}{1 - \bar{b}\rho} - \frac{\bar{a}(T)\rho^2}{1 + 2\bar{b}\rho - \bar{b}^2\rho^2} \tag{4}$$

where, $\overline{r} = R/\overline{W}$, where $\overline{W}$ represents the molar mass of the mixture, and $\overline{a}$ and $\overline{b}$ denote the attraction parameter and the mixture co-volume, respectively. These coefficients are determined using the Van der Waals mixing rules presented in [24].

The two-phase thermodynamic state is determined by minimizing a function with N_s variables, where N_s represents the number of chemical species, to minimize the Gibbs energy of the entire system. However, computing the exact thermodynamic equilibrium becomes computationally expensive as the number of species increases. To address this, Pelletier et al. [16] proposed an approximate formulation for two-phase equilibrium in multi-component systems that maintains constant complexity, regardless of the number of species at the interface. This approximation assumes that both the liquid and gas phases share the same chemical composition, meaning that for any species k, $Y_{k,l} = Y_{k,v} = Y_k$. Finally, the saturation densities, ρ_l^{SAT} and ρ_v^{SAT}, are determined through a tabulation of the saturation curves as described by Pelletier et al. [16].

The turbulent combustion model (referred here as Infinitely Fast Chemistry Model (IFCM)) is based on an infinitely fast relaxation to chemical equilibrium, coupled with a β-pdf function of the mixture fraction to account for flame/turbulence interaction [25]. The IFCM methane-oxygen equilibrium considered in the present work is based on seven species (CH_4, O_2, H_2O, CO_2, CO, OH and H_2).

4 Geometry and Meshes

Two different mesh refinements, denoted as $M0$ and $M1$, are considered in this study and shown in Fig. 1. Detailed properties of the meshes are provided in Tab. 3. The $M0$ mesh is primarily refined in the presumed flame region and at the lip, aiming for rapid convergence and providing an initial insight into the thermodynamic fields. The $M1$ mesh was enhanced based on the simulation results from $M0$, with refinements focused on the region where oxygen is in a liquid state to achieve a better-resolved dense core, and at the flame location. Additionally, the second part of the chamber features a mesh size Δ_x that is two times smaller compared to $M0$

5 Boundary Conditions

The outlet boundary condition consists in characteristic non-reflecting boundary condition [26], imposing a static pressure $P_s = 17$ bar. Mass flow, temperature and species mass fractions (see Table 1) are imposed at the inlet through a characteristic formalism. The near-wall resolution Y^+ ranges from 120 to 500 for $M1$. Therefore, for the chamber wall boundary condition, a coupled velocity-temperature law-of-the-wall [27] is considered, which has proven to be accurate for rocket engines where large temperature gradients exist at the walls ([28–30]). The wall temperature is

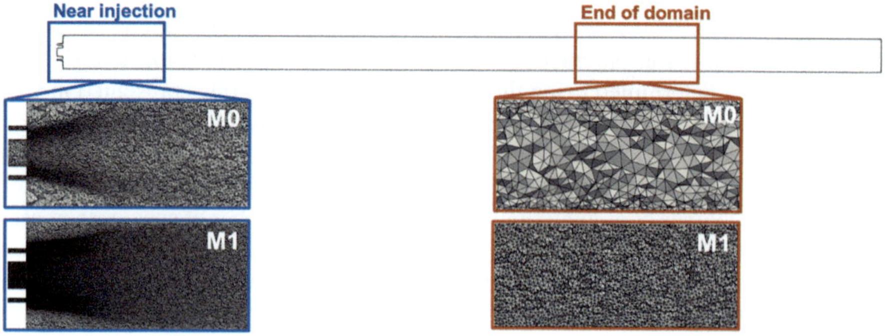

Fig. 1 Illustration of the meshes near the injection region and at the end of the domain

Table 3 Cell characteristic length (Δ_x) for the 2 meshes at various locations

Mesh	Injectors ($x < 0$ mm)	Flame ($x < 50$ mm)	End of domain (150 mm $< x$)	Cell number
M0	0.08 mm	0.2 mm	1 mm	$3.3.10^6$
M1	0.045 mm	0.07 mm	0.45 mm	$25.1.10^6$

imposed following the experimental profile, interpolated using the same third-order polynomial as in [18]. Other walls are treated as adiabatic with a standard wall law.

6 Results

The results of the present simulations are presented below and, when possible, compared to experimental data. Additionally, a comparison is made with the results obtained by [29] for the same operating point. It is noteworthy that the chemistry modeling and liquid injection methods differ between Blanchard's work [29] and the present study.

Temporal averages are performed over 3 convective times t_{conv}, where t_{conv} accounts for both the liquid and the gaseous convective times, respectively t_l and t_g, defined in Eq. (5):

$$t_l = \frac{L_l}{u_l} \tag{5a}$$

$$t_g = \frac{(L_{tot} - L_l)}{u_g} \tag{5b}$$

Table 4 Numerical cost per t_{conv} obtained on IRENE (GENCI)

Meshes	M0	M1
hCPU	42 000	170 000

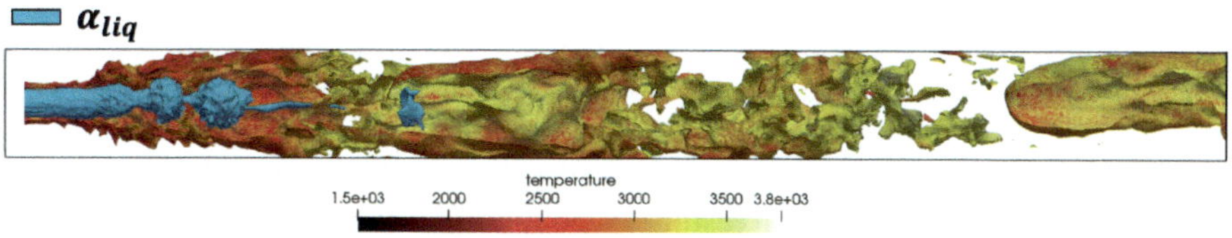

Fig. 2 Instantaneous 3D iso-surfaces of $\alpha_l = 0.5$ (blue) and mixture fraction $Z = Z_{st}$ colored by the temperature

with L_l the liquid core length based on $\alpha_l = 0.5$ iso-contour, and L_{tot} the length of the combustion chamber. The total convective time is then t_{conv}:

$$t_{conv} = t_l + t_g \tag{6}$$

The velocity in the cold region remains relatively constant within this iso-contour, with an average value of $u_l = 9$ m/s. Consequently, given $L_l = 42$ mm, an estimation of t_l is ≈ 5 ms. Additionally, considering $L_{tot} = 0.25$ m and $u_g = 260$ m/s, an approximation of t_g is $t_g \approx 1$ ms. Finally, $t_{conv} \approx 6$ ms.

As seen in Table 3, the liquid phase is located in the most refined region of the mesh. Due to the higher sound speed in the liquid phase, the explicit convective time step, constrained by the CFL condition, is reduced compared to the gaseous phase. The resulting time steps and CPU costs per convective time are provided in Table 4.

6.1 General Overview

A 3D visualization of the liquid core ($\alpha_l = 0.5$) and the flame location ($Z = Z_{st}$, with Z the mixture fraction) is presented in Fig. 2. The diffusion flame, located between the two propellant jets, expands radially before impinging on the wall. Subsequently, the flame closes, dividing the hot gases region from the fresh oxygen region. The cold liquid O2 jet progressively heats along the x-axis until it fully evaporates, marking the beginning of the burnt gas region. Considering the average contour at $\alpha_l = 0.5$ to represent the liquid core, its length is estimated to be approximately 18 mm in this case, compared to the 33 mm predicted by the correlation proposed by Woodward et al. [31]. Interestingly, the detachment of liquid structures, driven by shear forces and surface tension, can be observed in Fig. 2.

6.2 Wall Heat Fluxes and Pressure Profiles: Comparison with Experimental Data

The axial pressure along the chamber wall is shown in Fig. 3a for the current simulation, and compared with experimental data, as well as with Blanchard's [18, 29] results, who considered a more detailed chemistry. In Blanchard's simulation, the liquid core is modelled by a conical boundary condition with a length following Woodward's correlation [31], and Lagrangian particles are directly injected along this modeled liquid core, with assumed size-distribution. All meshes show very good agreement with the experimental results. Accounting for the liquid phase in our simulations significantly improves the pressure predictions near the injector compared to the results reported by [29].

Axial wall heat fluxes are shown in Fig. 3b, and compared to experimental data as well as to Blanchard results [18]. In contrast to the pressure profiles, significant differences are observed in the wall heat losses between $M0$ and $M1$. Indeed, a sudden increase in heat losses is observed at the axial position $x = 30$ mm for $M0$, while it is not observed for mesh $M1$. Additionally, the $M0$ heat fluxes reach a plateau closer to the injector compared to the other meshes, as well as to the experimental data and Blanchard's results. This behavior is attributed to the presence of a shorter flame. However, the $M1$ case shows very good agreement with experimental data within the initial 100 mm of the combustor, indicating that accounting for liquid injection significantly improves wall heat flux predictions in the first half of the combustor. Furthermore, the axial position of maximum heat loss is accurately captured compared to the experimental data. Beyond 100 mm, however, wall heat losses are slightly underestimated, with a discrepancy of about 10% for $x > 150$ mm.

This section has highlighted the importance of simulating the liquid injection compared to the Lagrangian injection model considered in Blanchard [18]. In the following section, averaged 2D fields are compared between our results on the mesh

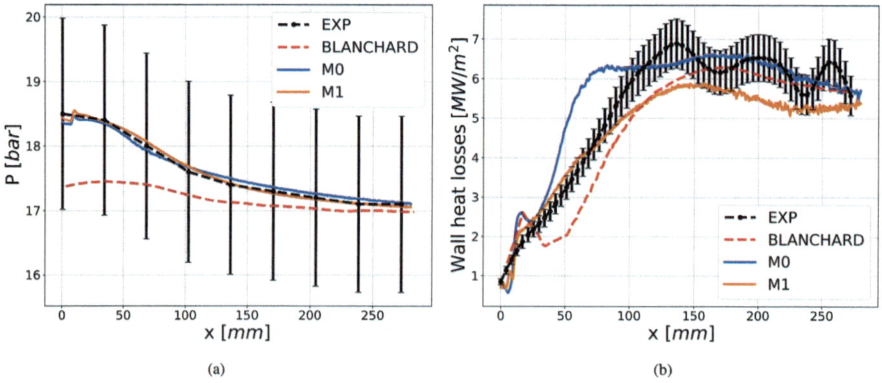

Fig. 3 Axial profiles of pressure (left) and wall heat flux (right) from experiment, [18] simulation and the present work

$M1$, and the simulation of Blanchard. It should be noted that meshes are similar between these two configurations.

6.3 Comparison with a Modeled Liquid Core Simulation

Conducting a comparative study with Blanchard's simulation [18, 29] is particularly interesting, as the configuration, operating point, equation of state, and numerical solver are identical. The key difference lies in the treatment of the liquid core: Blanchard's work models it, while the current study simulates it using a multi-fluid method. The time-averaged fields of temperature (top) and axial velocity (bottom) are illustrated in Fig. 4. The upper half shows Blanchard's results, while the lower half shows the $M1$ configuration. Although the colormap colors may slightly differ between the two cases, the colormap value boundaries are identical. For ease of visualization, these fields are scaled by a geometric factor of 0.5 in the x-direction.

The simulated liquid core, shown thanks to the iso-value $< \alpha_l >= 0.5$, exhibits a shorter length compared to the modeled counterpart. The cold region (blue) appears to be nearly two times shorter and thicker for the $M1$ case than the one predicted by Blanchard, as shown in Fig. 4a. This may due to evaporation processes occurring over too long distances for the Blanchard case, eventually due to too high Lagrangian particles axial velocity or/and diameter.

The initial radial flame expansion (for $0\,\text{mm} < x < 10\,\text{mm}$) reaches a higher axial and radial position than the Blanchard case. Moreover, in the Blanchard simulations,

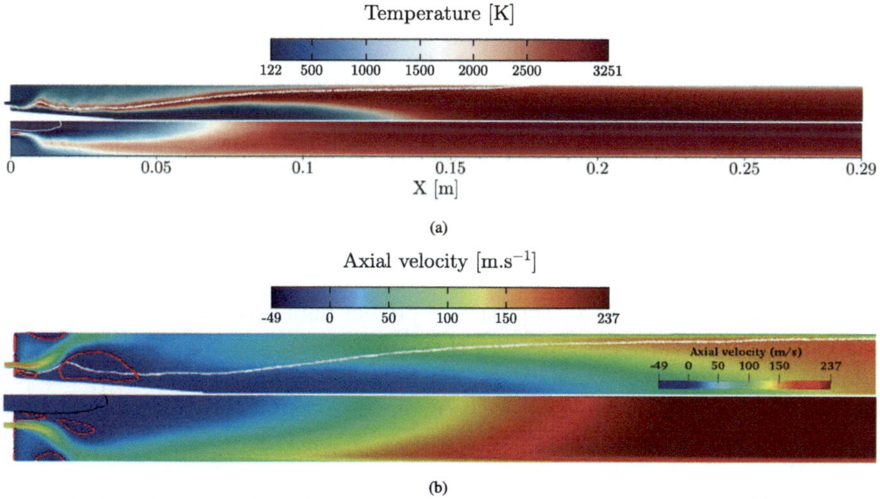

Fig. 4 Comparison of temperature (top) and axial velocity (bottom) average fields between the present work (lower half) and the simulation of Blanchard (upper half)

the high-temperature region contracts towards the center of the domain (for 10 mm $< x < 40$ mm), whereas our result shows a constant radial expansion of the hot front in this region. This can also be seen on the wall heat fluxes profiles between 10 and 40 mm, as depicted in Fig. 4b.

In the colder region, near the center of the domain (around x-axis), the axial velocity is lower when the liquid core is modeled. It must be noted that the velocity field in Blanchard's case only represents the gaseous velocity. Recirculation bubbles, defined by the iso-value $< u >= 0$ m/s (depicted as red lines in Fig. 4b), located at similar positions can be seen between the two computations. However, the recirculation bubble located between the CH4 and O2 jets is much larger in the case of Blanchard. Interestingly, in the Blanchard case, the flame tightens towards the center of the domain between $x = 10$ mm and $x = 30$ mm, occurring at the same location as the large recirculation bubble.

These comparisons demonstrate that accounting for the liquid core in the simulation can significantly alter the average flow topology compared to simulations using a pure Lagrangian injection. Furthermore, the flow topology in the vicinity of the injection region is high sensitivity to the model inputs. Additionally, there is a noticeable enhancement in the prediction of the average pressure distribution and wall heat flux, particularly in the first part of the combustion chamber where the liquid core is located.

7 Conclusions

In this paper, reactive simulations of a lab-scale liquid rocket engine configuration using a compressible multi-fluid method have been presented. The comparison with experimental profiles of wall heat flux and pressure reveals very encouraging results, especially in the vicinity of the injection region. Notably, a comparison with a purely Lagrangian injection emphasizes the importance of simulating the liquid core rather than modeling it. However, an observed underestimation of 10% in wall heat flux for $x > 100$ mm raises questions. Possible explanations for this underestimation include the equilibrium assumption of infinitely fast chemistry, leading to misestimations, particularly on the rich side. Another potential factor is that the law-of-the-wall might misestimate wall heat fluxes. These favorable outcomes underscore the capability of the proposed two-phase flows approach to predict liquid rocket combustion chambers within a Large Eddy Simulation (LES) framework, all while maintaining an affordable computational cost. In future works, a new mesh will be used to assess mesh convergence, and an investigation into the underestimation of the end chamber fluxes will be conducted.

Acknowledgements This work has been funded by CNES and ArianeGroup and was performed using HPC from CERFACS and benefited from GENCI-CINES HPC ressources (Grant 2024 - A0152B10157).

References

1. Silvestri, S., Celano, M.P., Kirchberger, C., Schlieben, G., Haidn, O., Knab, O.: Investigation on recess variation of a shear coax injector for a single element gox-gch4 combustion chamber. Transactions of the Japan Society For Aeronautical and Space Sciences, Aerospace Technology Japan, 14, 12 (2016)
2. Silvestri, S., Celano, M.P., Schlieben, G., Haidn, O.J.: Characterization of a multi-injector GOX/CH4 combustion chamber. In: 52nd AIAA/SAE/ASEE Joint Propulsion Conference. American Institute of Aeronautics and Astronautics (2016)
3. Celano, M.P., Silvestri, S., Schlieben, G., Kirchberger, C., Haidn, O.J., Knab, O.: Injector characterization for a gaseous oxygen-methane single element combustion chamber. In: Calabro, M., DeLuca, L., Frolov, S., Galfetti, L., Haidn, O. (eds.) Progress in Propulsion Physics. EDP Sciences (2016)
4. Hulka, J.R., Jones., G W.: Performance and stability analyses of rocket thrust chambers with oxygen/methane propellants. In: 46th AIAA/ASME/SAE/ASEE Joint Propulsion Conference, number M10-0882 (2010)
5. Von Sethe, C.B., Soller, S., Haidn, O.J.: Experimental investigation of a single element lox/gch4 heat-sink combustion chamber. 07 (2019)
6. Hardi, J., Deeken, J.C., Armbruster, W., Miene, Y., Haemisch, J., Martin, J., Suslov, D., Oschwald, M.: Lumen thrust chamber - injector design and stability analysis. In: 32nd International Symposium on Space Technology and Science (2019)
7. Fdida, N., Vingert, L., Mauriot, Y., Dorey, L.H., Théron, M.: Comparison of lox/methane and lox/hydrogen cryogenic spray combustion with simultaneous optical diagnostics. In: EUCASS, vol. 2019, pp. 1–10 (2019)
8. Lucien Vingert, Nicolas Fdida, Yves Mauriot, Ismael Ortega Colomer, Cornelia Irimea, David Delhaye, Ajmal Khan Mohamed, Vincent Corbas, and Marie Theron. Experimental investigations of liquid oxygen/methane combustion at very low mixture ratio at the mascotte test facility. In *32nd ISTS & 9th NSAT Joint Symposium*, 2019
9. Lechner, V., Betrancourt, C., Scouflaire, P., Vingert, L., Ducruix, S.: Dynamic characterization of wall temperature in lox/ch4 rocket engine operating conditions using phosphor thermometry. Proc. Combust. Inst. **39**(4), 5033–5041 (2023)
10. Hirt, C.W., Nichols, B.D.: Volume of fluid (vof) method for the dynamics of free boundaries. J. Comput. Phys. **39**, 201–225 (1981)
11. Sussman, M., Smereka, P., Osher, S.: A level set approach for computing solutions to incompressible two-phase flow. J. Comput. Phys. **114**, 146–159 (1994)
12. Sussman, M., Gerry Puckett, E.: A coupled level set and volume-of-fluid method for computing 3d and axisymmetric incompressible two-phase flows. J. Comput. Phys. **162**(2), 301–337 (2000)
13. Ménard, Thibault, Tanguy, Sebastien, Berlemont, Alain: Coupling level set/vof/ghost fluid methods: Validation and application to 3d simulation of the primary break-up of a liquid jet. Int. J. Multiph. Flow **33**(5), 510–524 (2007)
14. Herrmann, M.: Detailed numerical simulations of the primary atomization of a turbulent liquid jet in crossflow. J. Eng. Gas Turbines Power **1326**, (2010)
15. Zuzio, Davide, Estivalèzes, Jean-Luc., DiPierro, Bastien: An improved multiscale eulerian-lagrangian method for simulation of atomization process. Comput. Fluids **176**, 285–301 (2016)
16. Pelletier, M., Schmitt, T., Ducruix, S.: A multifluid taylor-galerkin methodology for the simulation of compressible multicomponent separate two-phase flows from subcritical to supercritical states. Comput. Fluids **206**, 104588 (2020)
17. Schmitt, T., Ducruix, S.: Evaluation of large-eddy simulation coupled with an homogeneous equilibrium model for the prediction of coaxial cryogenic flames under subcritical conditions. Aerospace **10**(2) (2023)
18. Blanchard, S., Cuenot, B.: Large eddy simulation of a single-injector lox/gch4 combustion chamber using analytically reduced chemistry. In: DFG-TRR-40 Summer Program 2019, vol. 1, pp. 69–86 (2019)

19. Schonfeld, Thilo, Rudgyard, Michael: Steady and unsteady flow simulations using the hybrid flow solver avbp. AIAA J. **37**(11), 1378–1385 (1999)
20. Gicquel, L.Y.M., Gourdain, N., Boussuge, J.-F., Deniau, H., Staffelbach, G., Wolf, P., Poinsot, T.: High performance parallel computing of flows in complex geometries. Comptes Rendus Mécanique **339**(2), 104–124 (2011). High Performance Computing
21. Colin, O., Rudgyard, M.: Development of high-order taylor-galerkin schemes for les. J. Comput. Phys. **162**, 338–371 (2000)
22. Franck, N., Ducros, F.: Subgrid-scale stress modelling based on the square of the velocity gradient tensor. Flow Turbul. Combust. **62**, 183–200, 09 (1999)
23. Peng, D-Y., Robinson, D.B.: A new two-constant equation of state. Ind. Eng. Chem. Fundam. **15**(1), 59–64 (1976)
24. Poling, B.E., Prausnitz, J.M., O'connell, J.P.: Properties of Gases and Liquids. McGraw-Hill Education (2001)
25. Schmitt, Thomas: Large-eddy simulations of the mascotte test cases operating at supercritical pressure. Flow Turbul. Combust. **105**(1), 159–189 (2020)
26. Poinsot, T.J., Lele, S.K.: Boundary conditions for direct simulations of compressible viscous flows. J. Comput. Phys. **101**(1), 104–129 (1992)
27. Cabrit, O., Nicoud, F.: Direct simulations for wall modeling of multicomponent reacting compressible turbulent flows. Phys. Fluids - PHYS FLUIDS **21**, 03 (2009)
28. Maestro, D., Cuenot, B., Selle, L.: Large eddy simulation of combustion and heat transfer in a single element GCH4/GOx rocket combustor. Flow, Turbulence and Combust. **103**(3), 699–730 (2019)
29. Blanchard., S.: Multi-physics large-eddy simulation of methane oxy-combustion in liquid rocket engines. Ph.D. thesis, Ecole doctorale MEGEP - Dynamique des fluides - CERFACS, 9 (2021)
30. Potier, L.: Large Eddy Simulation of the combustion and heat transfer in sub-critical rocket engines. Ph.D. thesis, Université de Toulouse (2018)
31. Woodward, R., Pal, S., Farhangi, S., Santoro, R.: Lox/gh2 shear coaxial injector atomization studies at large momentum flux ratios. In: 42nd AIAA/ASME/SAE/ASEE Joint Propulsion Conference & Exhibit, vol. 7, 07 (2006)

Experiments

Design of a Two-Phase Blowdown Test Facility

Charlie Westpfel, Pawel Ogrodniczak, and Martin T. White

Abstract The realisation of turbomachinery suitable for two-phase expansion in organic Rankine cycle (ORC) systems could unlock significant performance benefits. Several recent studies have explored two-phase turbomachinery flows using numerical simulations, however further development is hindered by a lack of any experimental data. This paper outlines a proposed two-phase blowdown facility for studying the high-speed two-phase expansion of molecularly complex organic fluids in stationary components. An overview of the test rig is presented, before a summary of the sizing and selection of key components, such as the working fluid, tanks, pump and heater is presented. A lumped-mass transient blowdown model is used to predict the conditions within the test rig during a representative blowdown run. Finally, steady-state non-equilibrium simulations are used to examine the expected test conditions.

Keywords Two-phase expansion · Experimental facilities · Test rig design

1 Introduction

Two-phase expansion has been proposed as a suitable way of improving the second-law efficiency of organic Rankine cycle (ORC) power systems for low-grade heat recovery applications. In such a system, the expander operates with two-phase inlet conditions and the expansion process takes place entirely, or partially, within the two-phase saturation region. This removes the well-known pinch-point constraint, enabling a larger recovery of heat and enhanced power output. However, a major hurdle is the identification of expanders capable of achieving two-phase expansion.

C. Westpfel (✉) · M. T. White
University of Sussex, Brighton, UK
e-mail: cw529@sussex.ac.uk

M. T. White
e-mail: martin.white@sussex.ac.uk

P. Ogrodniczak
City St George's, University of London, London, UK
e-mail: pawel.ogrodniczak@city.ac.uk

A. Giauque et al. (eds.), *Proceedings of the 5th International Seminar on Non-Ideal Compressible Fluid Dynamics for Propulsion and Power*, ERCOFTAC Series 32,
https://doi.org/10.1007/978-3-031-92695-2_14

Volumetric expanders have generally been considered to be the most suitable for two-phase operation [1, 2], but compared to their turbomachinery counter-parts typically achieve lower efficiencies and are better suited for relatively low volumetric-expansion ratios. As such, there has been a growing interest in two-phase expansion with turbomachinery [3–5]. This includes work of the current authors', which has focused on the development of the so-called *wet-to-dry* cycle, in which the two-phase expansion process of particular fluids could end up in the dry superheated vapour region. More specifically, it is proposed that if the wet-to-dry expansion process can be constrained to the stator vane of a turbine, then turbines could be exploited to achieve the benefits of two-phase expansion without any concerns around blade erosion. Studies have demonstrated the thermodynamic potential of the cycle [6], and studied the simulation of converging-diverging nozzles for wet-to-dry expansion accounting for non-equilibrium effects [7, 8]. However, experimental validation is lacking.

The investigation of single-phase NICFD flows in ORC turbines is being explored through a variety of experimental facilities. One such facility is the test rig for organic vapours (TROVA), which is an intermittent blow-down wind tunnel that utilises measurements of pressure and temperature alongside optical techniques for visualisation of vapour flows [9]. An additional facility is the ORCHID facility which is a continuous ORC test loop conceived to obtain measurement data of high-speed NICFD flows with a focus on ORC turboexpander technologies [10]. The closed loop organic wind tunnel (CLOWT) is another continuous test facility that is used to study NICFD organic vapour flows [11]. Additional facilities have also been constructed to explore the gas flow dynamics through similar principles to those described [12–14]. However, at present there have been no dedicated test facilities to study the high-speed two-phase expansion of organic fluids under conditions representative of passage flows within turbomachinery. Therefore, this paper details the preliminary design of such a facility to provide suitable experimental data to provide proof-of-concept and model validation. This paper provides details on the design and modelling of the proposed test facility and provides an outline of the defined component specifications.

2 Overview of Test Rig Concept

Due to the low-cost compared to that of a continuous facility, the blow-down setup will be implemented where the fluid is accelerated from a high to a low-pressure vessel due to a large pressure difference created during the charging of the system. The proposed setup for the facility can be seen in Fig. 1 where the key components are highlighted. The operating process is as follows: (1) two high-pressure liquid and vapour vessels (HPLV & HPVV) are charged with an organic fluid using a pump and isolated from the rest of the system and open to one another; (2) the HPLV and HPVV are heated using electric heaters on both tanks to increase temperature and pressure and achieve a two-phase state, and due to gravity liquid is decanted into the lower vessel (HPLV) leaving vapour in the upper vessel (HPLV). Individual

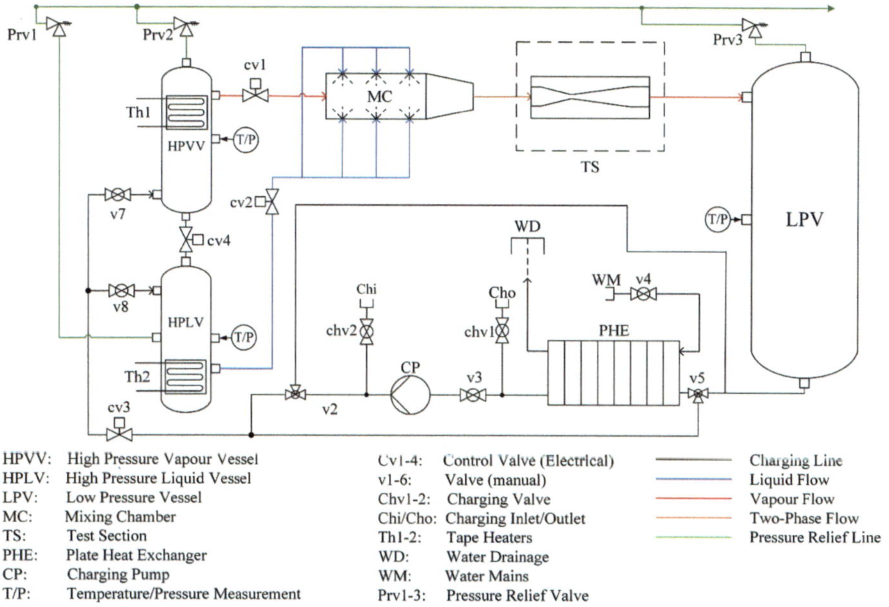

Fig. 1 Schematic of the proposed two-phase blowdown facility

pressure transducers and thermocouples mounted within each of the vessels provide measurements of the stagnation conditions throughout this heat addition process. (3) Once the target test conditions are achieved, the charging line is closed and the blowdown line is opened which is composed of the mixing chamber (MC) where gas is drawn from the HPVV and mixed with atomised liquid drawn from the HPLV. The pressure difference required for atomisation is obtained due to the drop in pressure in the HPVV as it discharges more rapidly than that of the HPLV. The system also possesses the ability to initially open the HPVV to drop its pressure prior to opening the HPLV line. The MC is followed by the test section (TS), namely a planar convergent-divergent nozzle, where the flow is expanded. The static pressure distribution within the nozzle will be recorded with pressure taps located along the wall of the converging-diverging test section. The test section will be designed to provide optical access to enable an initial qualitative visual inspection of the expansion process. (4) During the blowdown process fluid is transferred to the a low-pressure vessel (LPV) where the fluid is collected. (5) After a test run is complete, the charging process is repeated with the charging line consisting of a plate heat exchanger (PHE) where the fluid is condensed back to a liquid state and a charging pump (CP) to feed a set mass into the high-pressure vessels (HPLV & HPVV) and decrease the pressure within the low-pressure vessel.

3 Component Design and Sizing

3.1 Fluid Selection

Previous thermodynamic modelling of wet-to-dry power cycles have identified siloxanes, and most notably hexamethyldisiloxane (MM) and octamethyltrisiloxane (MDM), as the most suitable fluid candidates [6, 7]. These fluids exhibit a high degree of molecular complexity and have saturated-vapour lines with a positive gradient in the temperature-entropy (T-s) plane. However, whilst test facilities are in operation with siloxanes [9, 14, 15], siloxanes do pose some safety risks. Alternatively, Novec 649 has a saturation curve that facilities wet-to-dry expansion, and is also non-toxic and non-flammable [11, 16]. It also has a lower critical temperature, which means that wet-to-dry expansion can be achieved at lower temperatures. As such, whilst Novec 649 may not be the ideal fluid for a practical wet-to-dry cycle, it is a good choice for the planned test rig. However, the possibility of testing with MM or MDM within the proposed facility has been considered by ensuring all components with the rig, and the test cell, are ATEX rated.

In Fig. 2 and Table 1 the behaviour of Novec 649 is compared to MM and MDM, in which an ideal wet-to-dry cycle is found for each fluid by assuming a fixed ORC condensation temperature of 40 °C. A condensation temperature of 40 °C is selected to correspond to a realistic condensation temperature within a practical ORC loop, and does not correspond to the expected conditions within the LPV. With the condensation temperature defined, the corresponding saturated vapour entropy can be found. The ideal wet-to-dry expansion inlet condition then corresponds to an expansion from a saturated liquid with an entropy equal to this value; the ideal inlet conditions are shown in Fig. 2 and denoted in Table 1 as $T_{0,\mathrm{ideal}}$ and $P_{0,\mathrm{ideal}}$. This inlet condition is concurrent with the highest second-law efficiency that is possible using a wet-to-dry cycle operating with a heat-source temperature approximately equal to $T_{0,\mathrm{ideal}} + \Delta T_{\mathrm{pinch}}$, where the latter is an assumed pinch-point in the heat-addition heat exchanger [7]. An increase in the inlet pressure, and increase in the vapour quality, would result in a reduction in second-law efficiency, but widens the range of applicability of the cycle to other heat-source temperatures, and reduces the volu-

Table 1 Comparison of working fluids for the proposed test rig, including the critical properties (T_{cr} & P_{cr}), the ideal wet-to-dry cycle inlet conditions ($T_{0,\mathrm{ideal}}$ & $P_{0,\mathrm{ideal}}$) and the condensation pressure at 40 °C (P_{cond})

Fluid	T_{cr} [°C]	P_{cr} [bar]	$T_{0,\mathrm{ideal}}$ [°C]	$P_{0,\mathrm{ideal}}$ [bar]	P_{cond} [bar]	Flammable	Toxic
Novec 649	168.6	18.7	128.2	8.48	0.73	N	N
MM	245.5	19.3	164.9	4.79	0.11	Y (high)	Y (acute)
MDM	293.2	14.4	159.4	1.21	0.01	Y	Y (acute)

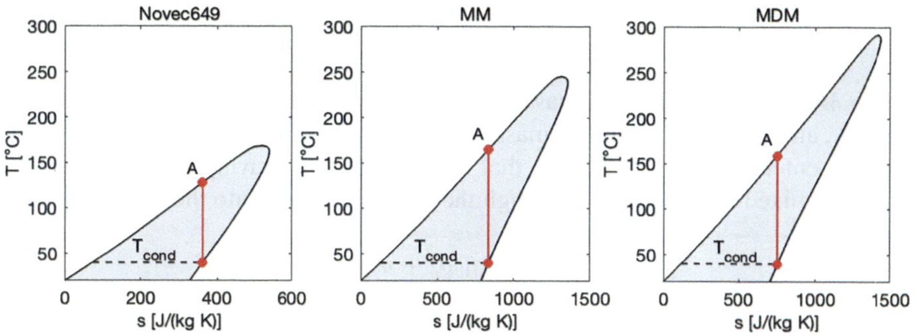

Fig. 2 Saturation domes for Novec 649, MM and MDM overlaid with the ideal wet-to-dry expansion process. Point 'A' corresponds to the ideal inlet condition, denoted $T_{0,\text{ideal}}$ and $P_{0,\text{ideal}}$. The condensation temperature is set to $T_{\text{cond}} = 40\ ^\circ\text{C}$

metric expansion ratio which could have practical benefits. For this simple analysis, the ideal wet-to-dry expansion process is assumed to be isentropic to determine the target tank stagnation conditions. However, an increase in entropy may aid in the transition from the two-phase domain to that of superheated vapour.

3.2 Tank Sizing

The three main tanks within the system need to be sized to provide the optimal blowdown conditions. Defining the optimal blowdown process is a trade-off between maximising the duration of the test and maximising the size of the test section, while keeping the total cost to within the allocated budget. To help with sizing the tanks, a simple transient lumped-mass model of the blowdown process has been constructed. The model considers each tank separately, and assumes there is no spatial variation of properties within each tank. Moreover, the two-phase mixture is modelled as a homogeneous two-phase mixture under thermal equilibrium. For the two high-pressure tanks, the mass and energy conservation laws can be written as:

$$\frac{\mathrm{d}M_i}{\mathrm{d}t} = -\dot{m}_i \tag{1}$$

$$\frac{\mathrm{d}(Mu)_i}{\mathrm{d}t} = -\dot{m}_i h_i \tag{2}$$

where M_i, u_i and h_i correspond to the mass, specific internal energy, and specific enthalpy of the fluid within a given tank, and $\dot{m}_i$ is the mass-flow rate leaving the tank. The conservation laws for the low pressure tank can be written as:

$$\frac{\mathrm{d}M_{\text{LPV}}}{\mathrm{d}t} = \dot{m}_{\text{LPV}} \tag{3}$$

$$\frac{\mathrm{d}(Mu)_{\mathrm{LPV}}}{\mathrm{d}t} = \dot{m}_{\mathrm{LPV}} h_0 \tag{4}$$

where M_{LPV} and u_{LPV} are the mass and specific internal energy of the fluid within the LPV, and $\dot{m}_{\mathrm{LPV}}$ and h_0 are the mass-flow rate and stagnation enthalpy of the fluid entering entering the LPV. Noting that the two streams leaving the high-pressure tanks are mixed before passing through the test section and into the LPV, it follows that:

$$\dot{m}_{\mathrm{LPV}} = \dot{m}_{\mathrm{HPLV}} + \dot{m}_{\mathrm{HPVV}} \tag{5}$$

$$h_0 = \frac{\dot{m}_{\mathrm{HPLV}} h_{\mathrm{HPLV}} + \dot{m}_{\mathrm{HPVV}} h_{\mathrm{HPVV}}}{\dot{m}_{\mathrm{LPV}}} \tag{6}$$

The two unknowns in this series of equations are the mass-flow rates leaving the two high-pressure tanks. The resolution of the mass-flow rates leaving the vapour and liquid vessels is determined iteratively. For an assumed static pressure within the mixing chamber, coupled with certain assumptions for the flow from upstream stagnation conditions in each tank (HPLV & HPVV) and assuming cross-sectional areas for the high-pressure vapour line and the total cross-sectional area of the atomisation nozzles, enables the calculation of mass-flow rate leaving the liquid and vapour vessels. More specifically, the flow from the HPVV to the mixing chamber inlet is assumed to be isentropic, while the flow from the HPLV to the mixing chamber is assumed to be isentropic up until the throat of the atomisation nozzle. Conservation of mass and energy sets the resulting properties within the mixing chamber, which then enables the determination of the mass-flow rate into the low-pressure vessel with the assumption of choked flow occurring in the test section. After the flow leaves the mixing chamber it is assumed that the flow is a two-phase mixture under thermal and mechanical equilibrium, and the flow from the mixing chamber outlet to the throat is assumed to be isentropic. The estimated pressure within the mixing chamber is iterated such that the difference between the flow rate entering the mixing chamber and leaving is minimised through non-linear optimisation. At present there is no method implemented with which to regulate the flows prior to the mixing chamber (i.e., pressure regulating control valve), and as such the vapour quality is unregulated and dictated by the stagnation conditions within each of the vessels, as well as the geometry of the mixing chamber and nozzle test section.

The testing time is defined as the duration for which the blowdown does not yield the formation of a shock within the test section geometry, and this is termed shock-free time. The shock-free time indicates the amount of time that occurs before a normal shock forms at the outlet of the nozzle, which is estimated by assuming homogeneous thermal equilibrium, meaning the shock relations can be easily solved. After this time, the normal shock would traverse along the nozzle from the outlet to the throat, before the flow becomes subsonic and then the blowdown process concludes. Maximising this time whilst remaining economically viable is paramount to achieve the target expansion conditions.

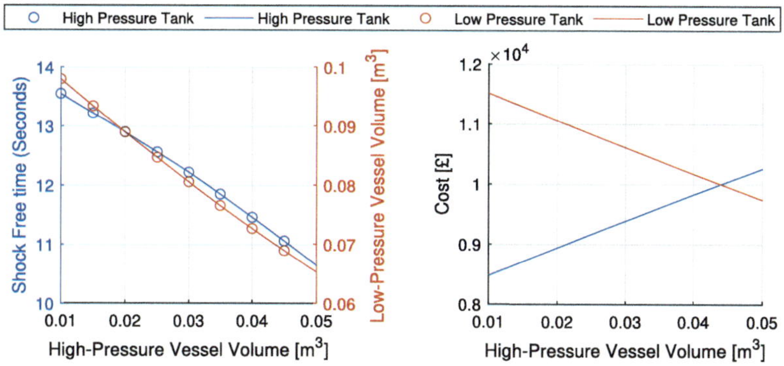

Fig. 3 Parametric study carried out with Novec 649 to explore resultant cost and shock-free time incurred through variation of high pressure vessel volume

A parametric study has been conducted to establish the volumes of the HPVV and LPV vessels that maximise the shock free time, while keeping a fixed HPLV volume of 10 L and a total tank budget of £20,000. The HPVV volume was increased in set increments where the cost was estimated from correlations from the chemical engineering industry, with the assumption of vertical tanks with length-to-diameter ratios of 5. The LPV volume is then determined from the remaining budget. The results from the parametric study, as reported in Fig. 3, indicate that a shock-free time of $\approx$13.5 s can be obtained for a HPVV volume of 10 L with a corresponding LPV volume of 100 L.

The evolution of a typical blowdown process, as predicted using the transient lumped-mass model is reported in Fig. 4. This corresponds to initial stagnation conditions within the high-pressure vessels of 10 bar and 135 °C, coupled with a pressure of 0.33 bar in the LPV. The mixing of the fluids takes place with liquid drawn from 45 nozzles of throat diameter 2 mm, and vapour is drawn through the vapour pipe which has a diameter 25.4 mm. The test section throat area is assumed to be 100 mm^2, whereas the nozzle outlet area is 417.5 mm^2. The design conditions for the converging-diverging nozzle, as described in Sect. 3.4, correspond to a nozzle inlet stagnation vapour quality of 0.3 and stagnation pressure of 6.7 bar. The design stagnation vapour quality is not taken from completely wet conditions to avoid very large volumetric expansion ratios associated with initiating expansion from a wet state. An inlet vapour quality of 0.3 has been suggested from prior numerical studies of the proposed expansion process, which indicates that at this vapour quality full vaporisation of the flow is possible after accounting for non-equilibrium effects [8]. The blowdown process shown in Fig. 4 demonstrates the transient nature of the facility and the range of nozzle inlet conditions expected. More specifically, between 1 and 13.7 s the nozzle inlet stagnation pressure drops from 9 bar to 6.6 bar, accompanied by a vapour quality increase from 0.23 to 0.43. It should be noted that the LPV vapour conditions reported correspond to those within the low-pressure vessel, and do not reflect the conditions at the outlet of the nozzle. Additionally, the initial

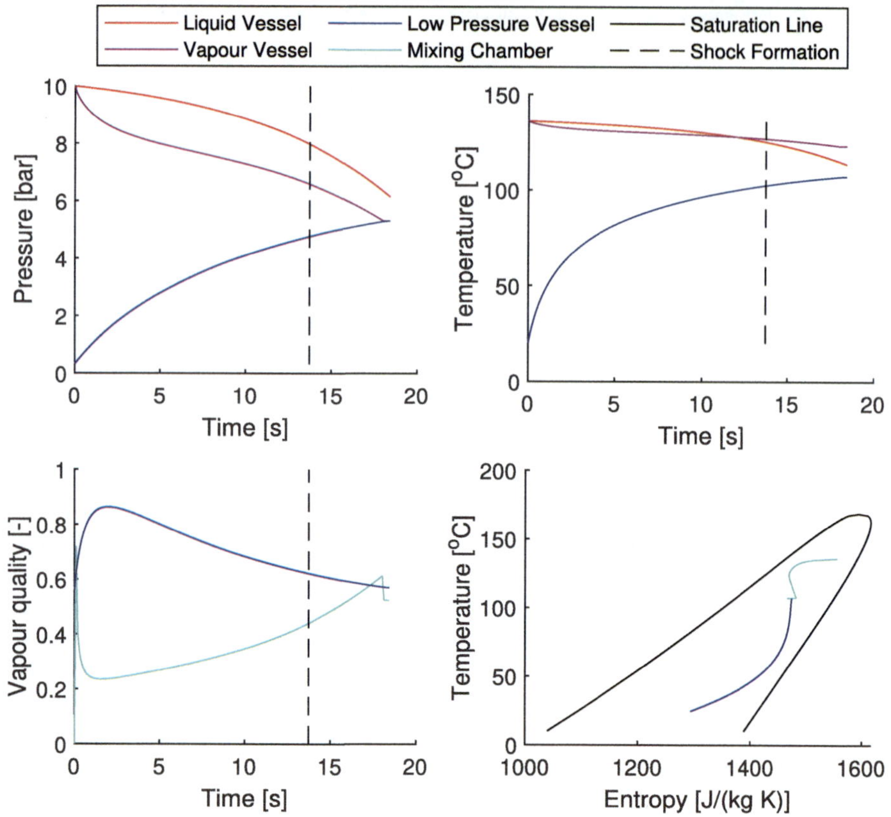

Fig. 4 Blowdown process for Novec 649 with a 10 L high-pressure vapour, 10 L high-pressure liquid vessel and a 100 L low-pressure vessel

mass discharge from the high-pressure vessels upon opening valves Cv1 and Cv2 (see Fig. 1) to fill the associated pipework has not been considered, thus the assumption is that there is no delay between initiation and the fluid reaching the LPV. It is acknowledged that this will reduce the experimental sampling time.

3.3 *Charging Line*

The intermittent nature of a blow-down facility requires that the high-pressure tanks be recharged. The charging process is a two-stage operation involving a pump to transfer mass between the tanks, followed by heating of the tanks to reach the target temperature and pressure. The pump is used to transfer the organic fluid from the LPV to the HPVV and HPLV, hence dropping the LPV pressure until the target mass in the high pressure vessels is achieved. The sizing of the pump is primarily defined by the charging time to ensure filling is not too rapid, thus sacrificing control and accuracy. The sizing of the pump was investigated through a separate transient lumped-mass model to monitor flow properties between the coupled high-pressure vessels and the LPV. Simulations indicate that a pump with a maximum flow rate of 0.6 L/min would complete the filling process in approximately 8 min. Decreasing the flow rate further would have no significant impact as it will only increase charging time. The heating of both the HPLV and HPVV during the charging process is responsible for the pressurisation of the high-pressure vessels and this is modelled by including a heat addition term in the energy conservation law, and assuming no losses to the surroundings. The required size of the heater, like that of the pump, is dictated by the target charging times. A parametric study indicates that a 1 kW electric heater mounted on each pressure vessel will be sufficient to heat Novec 649 to the target conditions, alongside the mass of the tanks, in under 2 h after the vessels have been charged with the required mass for the study.

3.4 Nozzle Design

The test section comprises of a planar converging-diverging nozzle, which is a simple representation of a supersonic stator vane within which full wet-to-dry expansion is hypothesised to occur. The nozzle for the test section is designed using an in-house inviscid non-equilibrium quasi-1D nozzle design tool previously developed to design nozzles for wet-to-dry expansion [7]. The design tool is based on an inverse-design approach whereby the target 1D pressure profile of the nozzle is defined as an input. The thermodynamic boundary conditions for this initial design are defined for an expansion starting from a vapour quality of 30%, rather than from saturated liquid. This case is selected to reduce the change in density across the nozzle, and such a boundary condition is arguably of more practical relevance than the ideal conditions defined in Sect. 3.1. The other design conditions correspond to an inlet stagnation temperature and pressure of 117.3 °C and 6.71 bar respectively, and a static outlet pressure of 0.61 bar. The nozzle is designed with a nozzle throat area of 100 mm^2, and the final nozzle has a throat height of 5.77 mm, a fixed passage width of 17.3 mm and a nozzle length of 173 mm. The performance of the nozzle has been assessed using a non-equilibrium CFD simulation. These simulations are based on a Eulerian-Eulerian two-fluid model with interfacial models to describe mass, momentum and energy exchange between the vapour phases and thus account for both thermal and mechanical equilibrium. Apart from the working fluid and boundary conditions, the setup for the CFD simulations is very similar to those described in the authors' previous study [8]. The results presented in Fig. 5 report the mass-flow averaged vapour quality and vapour volume fraction taken on planes at different streamwise locations, while contours of vapour mass-fraction and mixture velocity are reported in Fig. 6. The vapour quality is calculated from the ratio of the vapour mass-flux to the total mass-flux, while the mixture velocity is calculated based on the mass-fraction weighted sum of the liquid and vapour components. It should be noted that

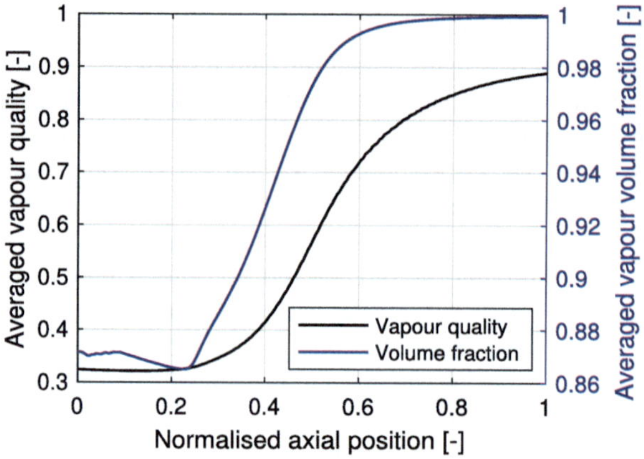

Fig. 5 Mass-flow averaged vapour quality and vapour volume fraction along the length of the nozzle, as predicted from the non-equilibrium CFD simulations

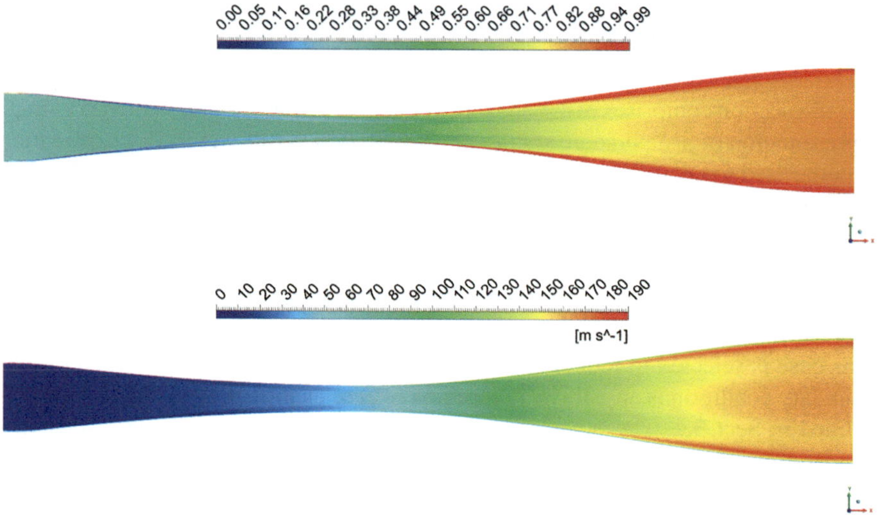

Fig. 6 Contours of vapour mass fraction (top) and mixture velocity (bottom)

the geometry shown is an initial design and this will be further optimised for the proposed test facility within future research.

4 Conclusions and Future Works

This paper has presented the conceptual design and preliminary component sizing for a two-phase blowdown test facility to study the wet-to-dry expansion of molecularly complex working fluids. The facility will initially operate with Novec 649, but is being designed to be suitable for siloxanes such as MM and MDM. The component sizing has been conducted using lumped-mass transient models of the pressure vessels and parametric studies have been conducted to determine the required tank sizes, charging pump size, and heating requirements based on the available budget. The final design is comprised of two 10 L tanks that, prior to initiating a blowdown run, will contain high pressure liquid and high pressure vapour respectively, alongside a single 100 L low-pressure tank. For these tank volumes, and an assumed nozzle area of 100 mm^2, simple estimates based on the assumption of homogeneous equilibrium between the phases indicate that shock-free isentropic expansion within the nozzle can be achieved for over 10 s.

References

1. Ottaviano, S., Poletto, C., Ancona, M.A., Melino, F.: Experimental investigation on micro-ORC system operating with partial evaporation and two–phase expansion. Energy Conver. Manag. **274**, 116415, 12 (2022)
2. Dawo, F., Buhr, J., Schifflechner, C., Wieland, C., Spliethoff, H.: Experimental assessment of an organic Rankine cycle with a partially evaporated working fluid. Appl. Thermal Eng. **221**, 119858, 2 (2023)
3. Tammone, C.: Two-phase expansion in turboexpanders for organic Rankine cycle power systems. Ph.D. thesis, Technical University of Denmark (2023)
4. Sampedro, E.O., Védie, L.: Wet to dry cycles for high temperature waste heat valorisation using a diphenyl-biphenyl oxide mixture. In: 6th International Seminar on ORC Power Systems, 11-13 October, Munich, Germany (2021)
5. Romei, A., Giostri, A., Spinelli, A.: Assessment of trilateral organic Rankine cycle for solar applications with innovative turboexpander concept. In: 7th International Seminar on Organic Rankine Cycle Power Systems, 4-6 September, Seville, Spain (2023)
6. White, M.T.: Cycle and turbine optimisation for an ORC operating with two-phase expansion. Appl. Thermal Eng. **192**, 116852, 6 (2021)
7. White, M.T.: Investigating the wet-to-dry expansion of organic fluids for power generation. Int. J. Heat Mass Transf. **192**, 122921, 8 (2022)
8. Ogrodniczak, P., White, M.T.: Non-equilibrium CFD simulation of the wet-to-dry expansion of the siloxane MM in a converging–diverging nozzle. Appl. Thermal Eng. **234**, 121158, 11 (2023)
9. Cammi, G., Conti, C.C., Spinelli, A., Guardone, A.: Experimental characterization of nozzle flow expansions of siloxane MM for ORC turbines applications. Energy **218**, 3 (2021)
10. Head, A.: Novel Experiments for the Investigation of Non-Ideal Compressible Fluid Dynamics The ORCHID and First Results of Optical Measurements. Ph.D. thesis, Technical University of Delft (2021)
11. Reinker, F., Wagner, R., Hake, L., aus der Wiesche, S.: High subsonic flow of an organic vapor past a circular cylinder. Exp. Fluids **62**, 1–16 (2021)
12. Baumgärtner, D., Otter, J.J., Wheeler, A.P.S.: The effect of isentropic exponent on transonic turbine performance. J. Turbomach. **142**(8), 081007, 7 (2020)

13. Robertson, M., Newton, P., Chen, T., Costall, A., Martinez-Botas, R.: Experimental and numerical study of supersonic non-ideal flows for organic Rankine cycle applications. J. Eng. Gas Turbines Power **142**, 8 (2020)
14. Uusitalo, A., Turunen-Saaresti, T., Honkatukia, J., Dhanasegaran, R.: Experimental study of small scale and high expansion ratio ORC for recovering high temperature waste heat. Energy **208**, 118321 (2020)
15. Spinelli, A., Cammi, G., Conti, C.C., Gallarini, S., Zocca, M., Cozzi, F., Gaetani, P., Dossena, V., Guardone, A.: Experimental observation and thermodynamic modeling of non-ideal expanding flows of siloxane MDM vapor for ORC applications. Energy **168**, 285–294 (2019)
16. Alshammari, F., Pesyridis, A.: Experimental study of organic Rankine cycle system and expander performance for heavy-duty diesel engine. Energy Conver. Mana. **199**, 111998 (2019)

Mixed Convection at Supercritical Pressures

Marko Draskic, Jerry Westerweel, and Rene Pecnik

Abstract Fluids display sharp, non-linear variations of thermodynamic properties when they are heated at a supercritical pressure. As such, near-pseudo-critical heat transfer is often characterized by large variations in density, leading to sharp near-wall accelerations or strong stratifications when buoyancy becomes dominant. We study the modulation of heat transfer and turbulence by non-negligible buoyancy in such property-variant flows, for the development of near-pseudo-critical heat exchangers for supercritical energy conversion systems. In particular, a liquid-like, horizontal base flow of carbon dioxide at 88.5 bar and 32.6 °C is considered, which is subjected to a vertical heat flux of up to 12.0 kW/m^2 at Reynolds numbers of up to $Re_{\mathrm{Dh}} \leq$ 10.000. Here, optical- and surface temperature measurements are used concurrently to evaluate the flow. Integratced visualizations of the flow field show the onset of strong stratifications with limited heating rates in the near-pseudo-critical region. During unstable stratification, the channel flow is dominated by the upward motion of thermal plumes. When the stratification is stable, any vertical motion and turbulence present in an equivalent neutrally buoyant flow is suppressed. As a result, wall heat is removed more effectively in the unstably stratified configuration than in a forced convective flow, whereas the opposite is true for a stably stratified flow. The difference in the perceived heat transfer between the considered configurations increases as buoyancy becomes more dominant.

Keywords Supercritical carbon dioxide · Stratification · Heat transfer · Shadowgraphy

M. Draskic (✉) · J. Westerweel · R. Pecnik
Process & Energy, ME, TU Delft, Delft, The Netherlands
e-mail: m.draskic@tudelft.nl

R. Pecnik
e-mail: r.pecnik@tudelft.nl

A. Giauque et al. (eds.), *Proceedings of the 5th International Seminar on Non-Ideal Compressible Fluid Dynamics for Propulsion and Power*, ERCOFTAC Series 32,
https://doi.org/10.1007/978-3-031-92695-2_15

1 Introduction

Beyond a fluid's critical point, the clear distinction between its liquid and gas phase disappears. When a fluid is heated at a supercritical pressure, it undergoes a bubble-free, boiling-like process. This pseudo-boiling leads to pronounced non-linear variations in thermodynamic properties, particularly near the pseudo-critical temperature (T_{pc}) where the specific heat capacity peaks. Several innovative energy conversion systems for the decarbonization of power and high-temperature industrial heat operate in the near-pseudo-critical region. The substantial variations in density that can be induced in the heat exchangers of these cycles, can lead to flows that are dominated by buoyant effects. Consequently, heat transfer in these systems is highly dependent on configuration and direction, with empirical rates varying significantly from ideal rates [1]. Unlike for systems with ideal fluids that follow the Oberbeck-Boussineq (OB) appromixation, this behavior does not require extreme temperature gradients but is instead induced at moderate heat transfer rates and temperature gradients at a supercritical pressure (Fig. 1).

The stratification type, determined by the direction of the vertically imposed density gradient, plays a critical role in heat transfer when buoyancy is non-negligible. In unstable stratification a lighter fluid is formed below denser fluid in a bottom-upward heated flow, creating warm plumes and cold downdrafts. Here, the increased upward transfer of momentum enhances turbulence and wall-fluid heat transfer compared to a forced convective flow [2, 3]. Conversely, in stable stratification, where lighter fluid is formed above a denser bulk during downward heating, vertical movement is suppressed, reducing turbulent mixing and heat transfer efficiency [4–6]. This suppression occurs because the potential energy required to displace fluid parcels vertically is unmet along a density gradient, leading to elongated thermal structures aligned with the flow direction and decreased near-wall turbulence [6–8].

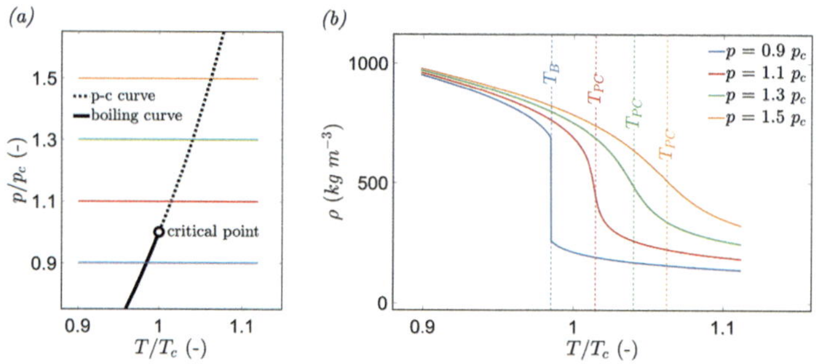

Fig. 1 The density of carbon dioxide, for which $T_c = 31.0°C$ and $p_c = 73.8$ bar, is shown in **b** for the isobaric heating processes shown in **a**. The pseudo-critical (p-c) curve is defined at the local maxima of the specific heat c_p. The p-c curve embodies the informal boundary between liquid-like carbon dioxide (to the left of the curve) with gas-like carbon dioxide (on the right)

The stable and unstable stratification of mixed convective flows is well-documented under the Oberbeck-Boussinesq approximation, or with linearly varying thermodynamic properties. However, discussions on the strong stratification with fluids with sharp, non-linear variations in thermodynamic properties are limited, particularly in an experimental context. It is however in systems with such fluids in particular, for which strong stratifications most readily prevail in practical applications.

In our work, horizontal flows of carbon dioxide at supercritical pressure with non-negligible buoyancy are experimentally explored, to predict and optimize the performance of heat exchangers of energy conversion systems at supercritical pressures. In particular, hydrodynamically developed horizontal channel flows to which one-sided vertical heating is imposed are considered. The channel can be heated in both bottom-up and top-down configurations, allowing the examination of both stable and unstable stratifications. In the our experimental facility, optical measurements are used concurrently with surface temperature measurements to study the flow at high temporal resolutions, overcoming the limitations of thermal inertia in the pressure vessel walls.

2 Methodology

The test section used to perform optical experiments with carbon dioxide in this work is shown schematically in Fig. 2. The test section is part of a larger experimental facility, in which carbon dioxide is circulated naturally in a vertically oriented closed flow loop. In the test section, bilateral optical access is provided to a rectangular channel using tempered borosilicate visors. At the streamwise location of the sight glasses, a unidirectional and constant vertical heat flux is applied to the channel. By inverting the test section, the heating can be applied in either vertical direction. In both configurations, the temperature of the heated surface is measured locally, using K-type thermocouple probes with a nominal uncertainty of $\pm 0.5°C$. The thermodynamic condition at the test section inlet is evaluated with a resistance

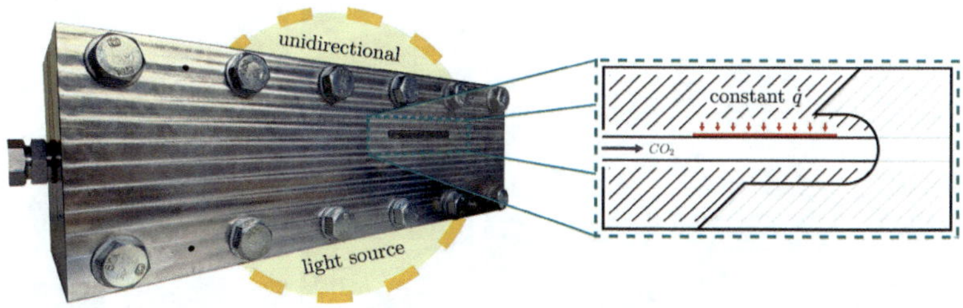

Fig. 2 Image of current test section (left) with schematic partial cross section of test section channel (right). A collimated light source is positioned behind the test section, creating a variable brightness shadowgram on the other side of the test section when the carbon dioxide is property-variant and turbulent. The rectangular test section channel is locally heated in a single vertical direction

thermometer bulk temperature measurement with a nominal accuracy of $\pm 0.1\,^{\circ}C$, and an absolute pressure measurement with a nominal uncertainty of ± 0.16 bar or 0.1%. The thermocouples are calibrated and shifted against the readings of the resistance transmitter in situ. The imposed steady state mass flow rate $\dot{m}$ is evaluated from the enthalpy difference over the heated vertical leg of the natural circulation facility. The uncertainty of the shown results is evaluated by assuming the independence of the respective variables measured by the transmitters in the system, allowing for the propagation of their respective errors.

A collimated-LED shadowgraphy setup is used to yield shadowgrams in this work. Here, a variable-brightness image is projected by an initially parallel light bundle after it passes the non-ideal carbon dioxide. Images are recorded using a Pixelink PL-D755MU-T CMOS camera (2448 × 2048 pixels, monochromatic). The exposure time of the sensor is maintained at 20 μs for all shadowgrams. Furthermore, the contrast of each image is stretched to cover the full brightness range.

3 Results

Instantaneous shadowgrams of an unheated, neutrally buoyant flow of carbon dioxide at supercritical carbon dioxide are shown in Fig. 3. The shadow cast on the image plane by the property-variant carbon dioxide reveals the presence of irregular thermal structures along the width of the channel. When the carbon dioxide is turbulent, its naturally present variations in refractive index show the motion of depth-integrated thermal structures in the shadowgram. As such, a comparison of the consecutive shadowgrams in Fig. 3 allows for the seedless visualization of the flow field [9, 10]. The resulting flow field is non-planar, as the refractive index is integrated across the channel. Therefore, its interpretability diminishes in the large Reynolds number limit; when many smaller thermal structures overlap laterally, or when structures

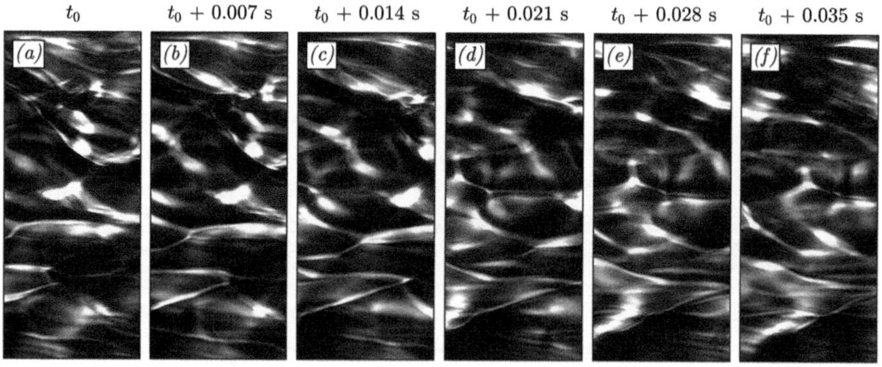

Fig. 3 Consecutive shadowgrams in a neutrally buoyant, unheated flow of carbon dioxide. Here, $Re_{Dh} = 3.8 \cdot 10^3$, $p = 88.5$ bar, $T_b = 32.6\,^{\circ}$C

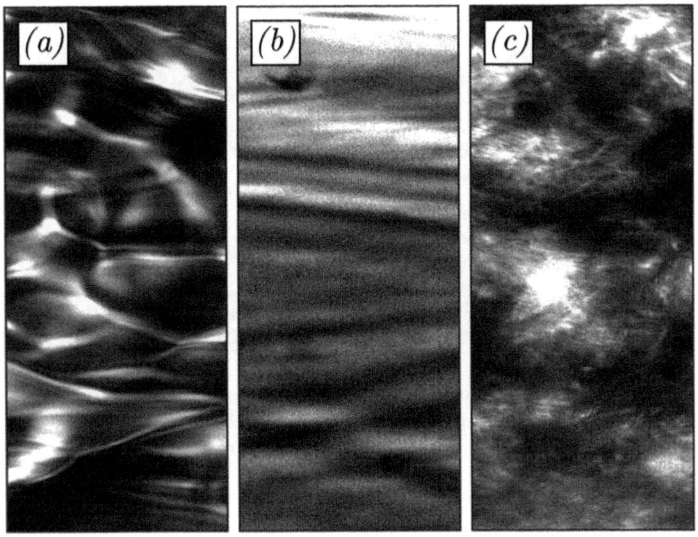

Fig. 4 Comparison of an unheated flow in **a**, a top-heated flow in **b**, and a bottom-heated flow in **c**. Here, $Re_{Dh} = 3.8 \cdot 10^2$, $p = 88.5$ bar, $T_b = 32.6$°C. The imposed wall-normal heating rate in both **b** and **c** is $\dot{q} = 12.0$ W/m^2

dissipate quickly. Furthermore, the reconstructed flow field undercaptures the actual fluid velocity in the low Reynolds number limit, as less inhomogeneous turbulent thermal tracer parcels are present to visualize the flow field with. Nevertheless, the current shadowgrams show distinguishable mean and instantaneous movement at all planes across the channel simultaneously for a band of Reynolds numbers.

Figure 4 shows a comparison between a neutrally buoyant, unheated flow, a flow that is heated in a stably stratified configuration, and a flow that is heated in an unstably stratified configuration. The thermodynamic condition and mass flow rate is constant for all three considered cases. Under strong top-down heating ($\dot{q} = 12.0$ W/m^2), the thermal structures originally present in the neutrally buoyant flow are elongated in the streamwise direction and disappear in the shadowgrams. Any vertical motion in the near-wall region is furthermore suppressed. As such, the resulting shadowgrams are mostly homogeneous in both space and time. When the flow is heated in a bottom-up setting ($\dot{q} = 12.0$ W/m^2), the optical signal is dominated by the upward movement of thermal plumes across the channel. Figure 5 shows the inception of thermal plumes after bottom-up heating is commenced for a laminar base flow. In the buoyancy dominant limit, the irregular and intermittent structures overlap laterally, and vertical movement is difficult to distinguish in the resulting blurry shadowgrams. Nevertheless, the upward momentum transfer is discernible under mixed convection and moderate heating rates within the considered range.

The considered shadowgrams show significantly different trends between the two considered stratification types. Under unstable stratification, increased momentum transfer is perceived, away from the heated wall towards the bulk of the flow. On

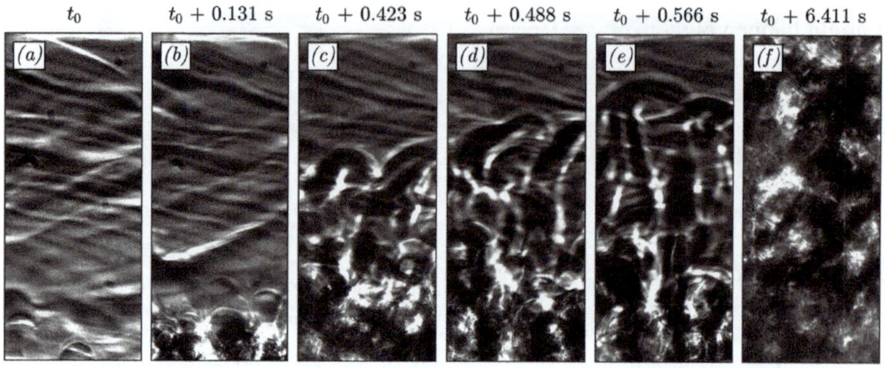

Fig. 5 Consecutive shadowgrams in an unstably stratified, bottom-heated flow of carbon dioxide show the upward ejection of low-density thermal plumes. Here, $Re_{Dh} = 1.0 \cdot 10^3$, $p = 88.5$ bar, $T_b = 32.6$°C, $\dot{q} = 12.0$ W/m^2. At $t = t_0$, the bottom-up heating is commenced

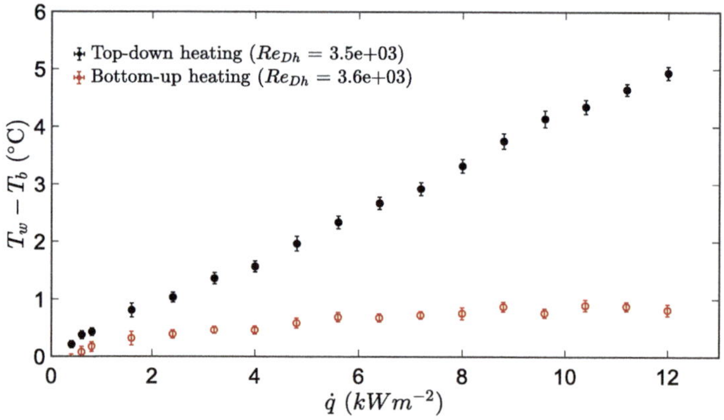

Fig. 6 The difference between wall- and bulk temperatures T_w and T_b for the cases shown in Fig. 4. The black data points show a stably stratified flow, whereas data of an unstably stratified flow is shown with red, unfilled markers

the contrary, when the flow is stably stratified, most vertical motion is suppressed. Figure 6 shows that the modulation in vertical momentum transfer in either stratification type results in an equivalent change of the heat transfer rate. At a constant heating rate for a buoyancy-dominated channel flow, heat is removed more effectively in the unstably stratified configuration. For the cases shown in Fig. 4, an up to five fold difference of the wall-temperature increase is found between the two considered vertical heating configurations. In an equivalent setting with an ideal fluid, the buoyant contribution would be negligible, and no significant difference in the heat transfer rates of the two considered configurations would be found.

4 Conclusions and Future Works

This work presents a preliminary optical investigation of a horizontal channel flow of carbon dioxide at supercritical pressure. The flow is heated vertically to incite both stable and unstable stratifications. Shadowgraphy is used to image fluid motion across the considered channel. Here, shadowgrams show increased upward motion in the unstably stratified configuration, in which buoyancy is non-negligible. As the vertical momentum transfer rate increases, the wall-normal heat transfer rate increases accordingly for the parameter range considered in this work. When the flow is heated in a stably-stratified configuration, both wall-normal motion and turbulent structures are suppressed in the near-wall region. As a result, the heat removal rates in the unstably stratified configuration far exceed those in a stably stratified arrangement.

References

1. Wang, Q., Jinliang, X., Zhang, C., Hao, B., Cheng, L.: A critical review on heat transfer of supercritical fluids. Heat Transf. Eng. **44**(21–22), 1969–1994 (2023)
2. Garai, A., Kleissl, J., Sarkar, S.: Flow and heat transfer in convectively unstable turbulent channel flow with solid-wall heat conduction. J. Fluid Mech. **757**, 57–81 (2014)
3. Pirozzoli, S., Bernardini, M., Verzicco, R., Orlandi, P.: Mixed convection in turbulent channels with unstable stratification. J. Fluid Mech. **821**, 482–516 (2017)
4. Ohya, Y., Neff, D.E., Meroney, R.N.: Turbulence structure in a stratified boundary layer under stable conditions. Boundary-Layer Meteorol. **83**, 139–162 (1997)
5. García-Villalba, M., Del Alamo, J.C.: Turbulence modification by stable stratification in channel flow. Phys. Fluids **23**(4) (2011)
6. Williams, O., Hohman, T., Van Buren, T., Bou-Zeid, E., Smits, A.J.: The effect of stable thermal stratification on turbulent boundary layer statistics. J. Fluid Mech. **812**, 1039–1075 (2017)
7. Smith, K.M., Caulfield, C.P., Taylor, J.R.: Turbulence in forced stratified shear flows. J. Fluid Mech. **910**, A42 (2021)
8. Zonta, F., Sichani, P.H., Soldati, A.: Interaction between thermal stratification and turbulence in channel flow. J. Fluid Mech. **945**, A3 (2022)
9. Jonassen, D.R., Settles, G.S., Tronosky, M.D.: Schlieren "piv" for turbulent flows. Opt. Lasers Eng. **44**(3–4), 190–207 (2006)
10. Hargather, M.J., Lawson, M.J., Settles, G.S., Weinstein, L.M.: Seedless velocimetry measurements by schlieren image velocimetry. AIAA J **49**(3), 611–620 (2011)

Rarefaction Shockwave Experiments in Dense Organic Vapours

Chandrasekhar Medipati, Nitish Chandrasekaran, Chiara Falsetti, and Piero Colonna

Abstract We experimentally investigate the nature of propagation of rarefaction waves in the dense vapors of the organic fluid dodecamethylcyclohexasiloxane, D_6. The experiments were performed in the Asymmetric Shock Tube for Experiments on Rarefaction Waves (ASTER) facility that is designed to withstand pressures and temperatures of up to 15 bar and 400 °C, respectively. Recent improvements to the test facility allow the accurate control of fluid temperature and pressure during heating, cooling, and the expansion process. To ensure temperature homogeneity of the working fluid during the isochoric heating process, temperature measurements were performed simultaneously at various locations in the axial direction of the shock tube and under different heating conditions. The propagation velocity of the rarefaction waves was measured at various thermodynamic states of the fluid by employing the time of flight method. The deviation of the fluid behavior from classical gasdynamic behaviour was observed with repeatability when the thermodynamic state of the fluid was close to its thermal stability limit.

Keywords Shock tube · Rarefaction shock wave · Time of flight · Fundamental derivative of gasdynamics

C. Medipati (✉) · N. Chandrasekaran · C. Falsetti · P. Colonna
Propulsion and Power, Delft University of Technology,
Delft, The Netherlands
e-mail: m.r.medipati@tudelft.nl

N. Chandrasekaran
e-mail: n.b.chandrasekaran@tudelft.nl

C. Falsetti
e-mail: c.falsetti@tudelft.nl

P. Colonna
e-mail: p.colonna@tudelft.nl

A. Giauque et al. (eds.), *Proceedings of the 5th International Seminar on Non-Ideal Compressible Fluid Dynamics for Propulsion and Power*, ERCOFTAC Series 32,
https://doi.org/10.1007/978-3-031-92695-2_16

1 Introduction

Single-phase non-reacting fluids whose thermodynamic states are close to the vapor-liquid saturation and to the vapor-liquid critical point exhibit fluid dynamic behavior that is strongly different from that of conventional gasdynamics [1]. The region in which these gasdynamic behavior occurs is generally referred as nonclassical gasdynamic region, and it is a subpart of the Nonideal Compressible Fluid Dynamic (NICFD) region. These regions in a $P - v$ plane are characterized by quantifying the thermodynamic quantity called the fundamental derivative of gasdynamics (Γ—curvature of an isentrope in a $P - v$ plane) [2]. As depicted in Fig. 1a, if the value of Γ is positive ($\Gamma > 0$, classical), the nature of fluid dynamic behavior resembles that of ideal gasdynamics/ textbook gasdynamics, with the formation of compression shocks and expansion waves (RW). In contrast, if Γ <0 (nonclassical), the behavior of fluids differs qualitatively and quantitatively, with the formation of rarefaction shock waves (RSW) and compression waves (Fig. 1b). Notably, the extent as well as existence of $\Gamma < 0$ regions is a strong function of the molecular complexity and the thermodynamic state of the working fluid [3]. The higher the molecular complexity of the fluid, the region of occurrence of Γ <0 is relatively bigger [4]. The fluids exhibiting this unconventional behavior are usually referred to as Bethe-Zel'dovich-Thompson (BZT) fluids, named after the scientists who first proposed the existence of such behavior.

Over the decades, a large number of theoretical [2, 5] and numerical studies [3, 6] have claimed the existence of nonclassical gasdynamic effects. Moreover, several experimental investigations have been carried out to demonstrate the occurrence of rarefaction shock waves, but with no success [7–9], as the region where nonclassical effects occur is quite limited and close to the critical point where the fluid's thermal decomposition is also expected [10]. Hence, providing compelling experimental proof of this fluid dynamic behavior poses several challenges. Current thermodynamic models, as well as numerical studies, predict that the size of the $\Gamma < 0$ region, in terms of the range of pressures and temperatures, is larger in siloxane D_6 when compared with other fluids of the same family [3, 6]. For this reason, D_6 is chosen for the current investigation to experimentally demonstrate the occurrence of nonclassical effects.

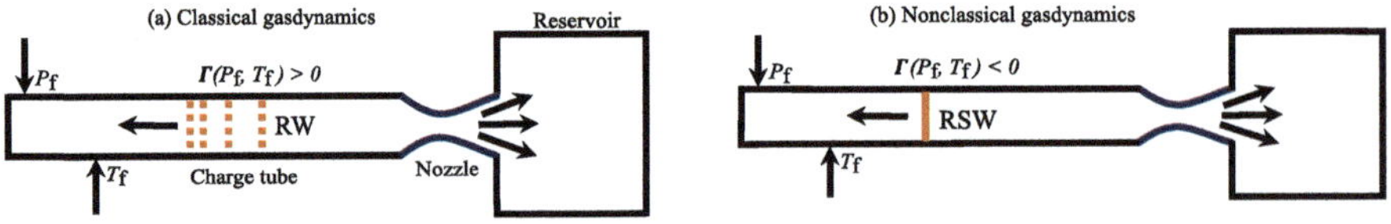

Fig. 1 Schematic to illustrate the difference in the fluid dynamic behavior between **a** classical and **b** nonclassical regimes of BZT fluids as they expand in a shock tube. P_f, T_f, and Γ denote the pressure, temperature, and fundamental derivative of gasdynamics of the working fluid. RW and RSW represent the rarefaction waves and rarefaction shock wave, respectively

In order to show nonclassical effects under various thermodynamic conditions, the same research group conducted a series of experiments to measure the wave speed utilizing a previous set-up, the Flexible Asymmetric Shock Tube (FAST), with siloxane (D_6) as the working fluid [8]. The attempts were unsuccessful due to the presence of axial temperature gradients on the shock tube wall. Since Γ is sensitive to temperature gradients, which has altered the local wave propagation behavior significantly [8, 11]. A new shock tube facility, the Asymmetric Shock Tube for Experiments on Rarefaction Waves (ASTER), was commissioned recently to demonstrate the existence of rarefaction shock waves [12]. The commissioning and the preliminary wave speed measurements indicating the presence of nonclassical effects have been reported in a recent study. For example, a constant wave speed with pressure drop has been observed, in contrast to a monotonic decrease if the fluid is in an ideal thermodynamic state [13]. However, a thorough investigation with precise initial experimental conditions is lacking from our recent study, which aids in proving the existence of non-classical effects with repeatability.

Hence, the main objective of the present study is to determine the initial thermodynamic state of the fluid more accurately by ensuring the temperature gradients in the fluid are absent. This helps us to prove the existence of non-classical effects with repeatability. Also, the possibilities of thermal degradation of the fluid when stressed at extremely high temperatures over a long period of time are discussed [14]. To address these issues, a few simple yet fruitful modifications are employed to the existing facility (ASTER), as discussed in the next section.

2 Experimental Facility

The experimental facility, working fluid, and experimental methodology are detailed in [15]. Hence, only the key details and the modifications implemented in the existing facility are presented here.

Rarefaction shock wave (RSW) experiments are conducted in the ASTER shock tube facility, and its schematic is illustrated in Fig. 2. It consists of a charge tube (CT), a Nozzle, a Low-pressure tank (LPT), a Glass-silk heating jackets, a Vacuum pump, a Manual pump, Pressure transducers (static and dynamic), and Thermocouples (for both CT wall and fluid temperature). Based on previous experimental results, a key modification employed here is introducing a thermocouple T_f (type - K) in the charge tube (CT). The thermocouple is flush mounted at 1.6 m from the nozzle throat to measure the temperature of the working fluid during the heating, cooling, and expansion processes. This directly measures the working fluid temperature, which is an important aspect when dealing with nonclassical effects. This aids us in estimating the value of Γ using the model with better accuracy. All the thermocouples were calibrated using a PRESYS dry block calibrator (T-350P) with an accuracy of 0.1 °C. The thermocouple (T_f), in combination with a static pressure transducer (P_f) on the CT, allows to achieve and control the desired thermodynamic state of the fluid with repeatability (within 0.5%). Closely monitoring the temporal behavior of the working

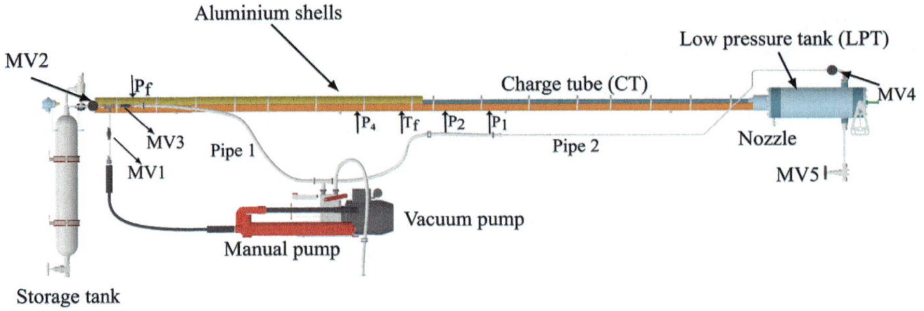

Fig. 2 Schematic of the ASTER facility

fluid's pressure and temperature also enables to predict any thermal decomposition of fluid during steady-state high-temperature heating conditions.

The main steps in the experimental procedure are: (i) Inserting a glass disk in the rear end of the CT, which acts as a diaphragm by separating the high-pressure (CT) and low-pressure (LPT) chambers during the heating process; (ii) Evacuating (to 30 mbar) the entire facility using the vacuum pump (by closing MV1 and MV5 and opening the remaining valves) about 3 times to ensure no traces of air/gases are present inside the ASTER. (iii) After closing MV3, the desired mass or volume of the working fluid is injected into the CT by operating the manual pump and opening MV1 & MV2. (iv) Finally, the required thermodynamic state is achieved by supplying the thermal energy to the fluid isochorically in the CT using the heating jackets (covering the CT).

In the current test facility, the final thermodynamic state is dictated by the initial amount of fluid pumped into the CT (as decided by the required P_f and T_f) and the set temperature of the heating jackets. Once the fluid attains a desired steady thermodynamic state, the glass disk is broken using a breaking pin mechanism [12], and the rarefaction waves propagate into the high-pressure CT. The dynamic motion of these waves was captured using the Kulite pressure transducers, which are placed on the CT wall. These high-frequency pressure measurements were used to analyze the wave speed, as discussed in detail in our earlier studies [8, 12]. The static and Kulite pressure transducers have a specified accuracy of ±0.1% over their full range of 16 bar and 21 bar, respectively. The working fluid used in the present investigation is D_6, which is 97.4% pure, as reported by gas-chromatography analysis. The relevant thermophysical properties of the fluid at the critical point are shown in Table 1.

Table 1 Thermophysical properties of D_6. MW is the molecular weight, and P_c, T_c, and v_c represent pressure, temperature, and specific volume at the critical point, respectively

Fluid	MW (kg/kmol)	P_c (bar)	T_c (°C)	v_c (m^3/kg)	Purity (%)
D_6	444.92	9.61	372.6	0.0036	97.4

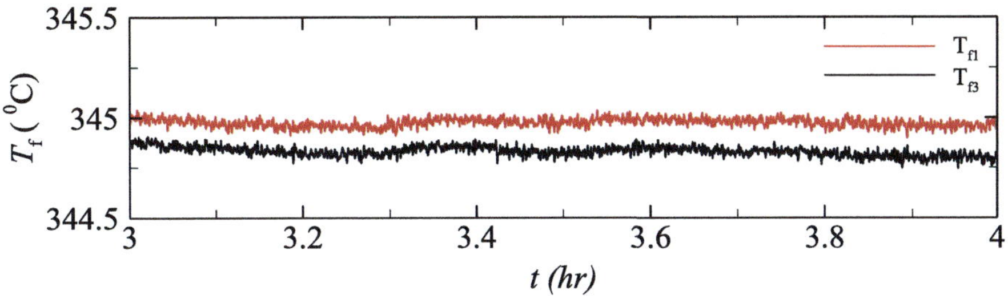

Fig. 3 Temporal variation of D_6 temperature during the isochoric heating process at two stations (1 and 3)

2.1 *Facility Characterization*

Rigorous leak tests were carried out in sub-atmospheric and super-atmospheric conditions by monitoring the change in pressure (ΔP) of the CT volume ($V \sim 0.0022\,\mathrm{m}^3$) over a period of 20 h (Δt), respectively. The corresponding leakrates ($\mathrm{LR} = V \cdot \Delta P \cdot \Delta t^{-1}$) were estimated to be 5×10^{-5} and 5.5×10^{-4} mbar $\cdot\, l \cdot s^{-1}$, respectively. As stated earlier, RSW experiments are very sensitive to the initial thermodynamic state of the fluid and the temperature gradient in the axial direction of the fluid[8, 11]. Therefore, before performing RSW experiments, simultaneous temperature measurements at different stations, as well as thermodynamic states of the fluid, were conducted. To accomplish this, two thermocouples (type-K) at various stations of the CT were installed by replacing the existing Kulites to record the fluid temperature during the isochoric heating process of D_6. Figure 3 depicts the temporal variation of temperature during 1 h at stations 1 and 3 when the fluid was heated steadily at about 345 °C (0.925 T_c). The trends highlight that the fluid temperature at both stations is steady, i.e. the standard deviation is within 0.01 % of its mean value. Also, the temperature difference between the stations falls within the uncertainty limit of the thermocouples (0.3 °C). This ensures that the fluid is fully homogeneous (free from axial temperature gradients), an important factor to be ensured before conducting RSW experiments.

2.1.1 Experimental Conditions

A series of experiments were conducted in the classical and nonclassical regimes of D_6 to demonstrate the nonclassical effects. The experimental conditions used for these investigations are shown in Table 2. Also, these are depicted in the T_r-s_r diagram of D_6 (Fig. 4). The key thermodynamic parameter in these tests is the value of Γ, which was estimated using the multiparameter EoS model for D_6 [16]. It is noteworthy to mention that there is a large uncertainty involved in the accurate prediction of the saturated vapor line at higher reduced temperatures and pressures as well as the critical point of the fluid using the existing model [13].

Table 2 Experimental conditions of the present study. P_f and T_f represent the pressure and temperature of the fluid in the CT. Γ is the fundamental derivative of gasdynamics predicted from the model

Test No.	P_f (bar)	T_f (°C)	Γ
1	3.6	350	0.82
2	9.35	371	−0.045
3	5	345	0.45
4	2.7	300	0.72

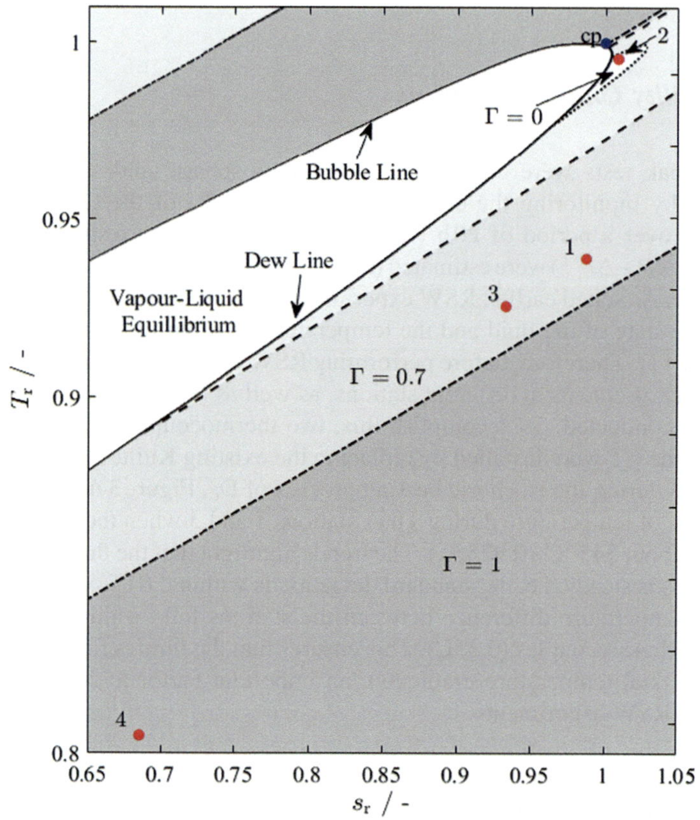

Fig. 4 $T_r - s_r$ diagram of D_6. Thermodynamic states at which rarefaction tests were performed are indicated with red dots. The critical point is marked with a blue colored dot. Few iso-Γ lines are also indicated. Data are calculated with the multiparameter EoS model for D_6 [16]

3 Results and Discussion

Once the fluid in the CT reached the desired thermodynamic state, RSW experiments were performed by rupturing the glass disk. The propagation of rarefaction waves was recorded using the Kulites connected to the CT wall. The sample pressure traces at stations 1, 2, and 4 for the experimental condition corresponding to test case 2 ($\Gamma = -0.045$), as indicated in Table 2, are depicted in Fig. 5a. It is evident from these traces that there is an abrupt drop in pressure during the expansion process of dense vapor, as anticipated. Using these pressure signals, by employing the Time of Flight method (ToF) [8, 12], the rarefaction wave speed (w) is quantified. This quantification needs the known separated distance between Kulite stations (L_T) along with the time taken by the wave (Δt) to travel from one Kulite station to the other. The equation used to calculate w is shown below:

$$w = \frac{L_T}{\Delta t} \tag{1}$$

The measured wave speed versus pressure drop (ΔP) between stations 1 and 4, during two repeated tests, is shown as test A and test B in Fig. 5b. It is observed that there is a region of constant wave speed (w_{14}) with a drop in pressure (between -0.5 and -1 bar), indicating the existence of nonclassical effects by the fluid as governed by its local thermodynamic state (Γ). This contrasts a monotonic decrease in wave speed with pressure drop if the fluid behavior/state is ideal [8, 12]. As it is possible to observe, the experimental result of test B shows a similar wave speed trend; see green curve in Fig. 5b. Both test A and test B are in close agreement with a negligible difference in wave speed (with a relative uncertainty 0.5%). This emphasizes the repeatability of the experiment and the deviation from the classical gasdynamic behavior of the fluid, indicating the existence of nonclassical effects.

However, an important aspect observed during these tests was a gradual increase in pressure with time repeatedly, especially under steady higher temperature heating

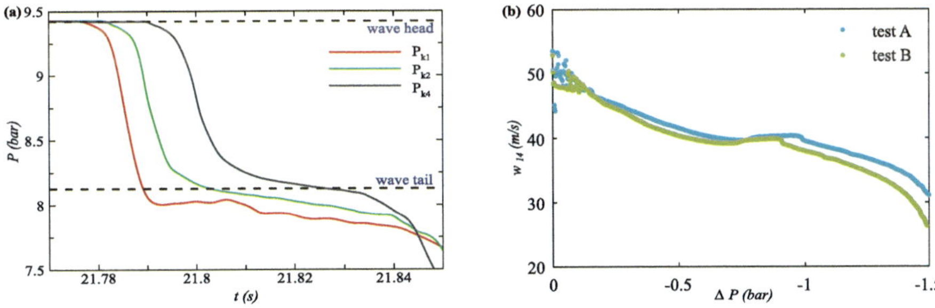

Fig. 5 **a** Sample pressure traces of Kulites at stations 1, 2, and 4 during RSW experiments (test 2). **b** Wave speed measurements in D_6 using the ToF method. The analysis is carried out within the limits as highlighted in plot **a** with dashed black lines

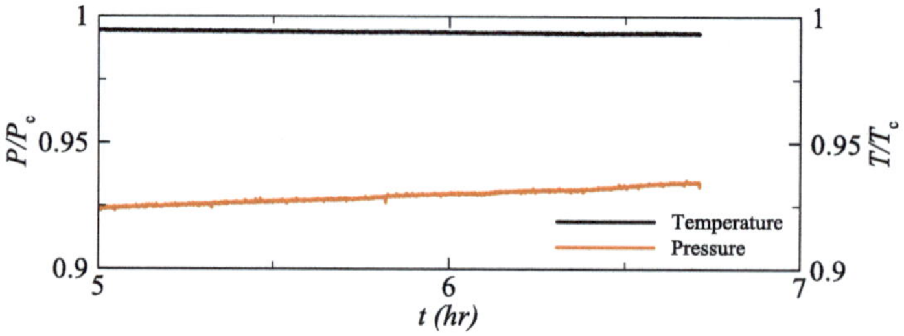

Fig. 6 Temporal variation of fluid temperature and pressure during the steady isochoric heating process. Pressure and Temperature are normalized with their corresponding critical properties

conditions (before rupturing the glass disk). A sample pressure and temperature traces corresponding to test 2, as shown in Table 2, are depicted in Fig. 6 for 2 h of the steady isochoric heating process. It is observed that there is a considerable increase in pressure of about 1.5% when the temporal change in temperature is negligible. This might indicate the thermal decomposition of the fluid since the test conditions are close to the limit of the thermal stability limit of the working fluid [10]. A similar signature in pressure traces during isothermal heating processes was observed in the existing literature [14, 17] while investigating the thermal stability limit of heat transfer fluids.

To conclude further, different sets of heating and cooling tests were conducted at different thermodynamic states of the fluid without rupturing the glass disk. The simultaneous measurement of fluid pressure and temperature allows to obtain the saturation curve on the lnP-$1/T$ plane during the heating and cooling processes, as shown in Fig. 7. The experimental conditions for this investigation correspond to test number 2, as shown in Table 2. Figure 7 illustrates the paths followed by the fluid during the isochoric heating and cooling processes as shown with red and green colored curves, respectively.

It is evident from Fig. 7 that there is a strong departure between the heating and cooling curves. Since the saturation pressure of the fluid during the cooling phase is higher, this may indicate the formation of lighter gases during the fluid's thermal decomposition, which was not observed in lower-temperature heating conditions. The formation of lighter gases would increase the speed of sound and decrease the strength of nonclassical effects. Hence, to confirm the thermal decomposition of the fluid at these thermodynamic states (close to critical state) and its impact on the wave speed measurements, rigorous gas chromatography analysis (GC) after every test has to be carried out. This is part of the ongoing research within the ASTER set-up, and it will be presented in a future study.

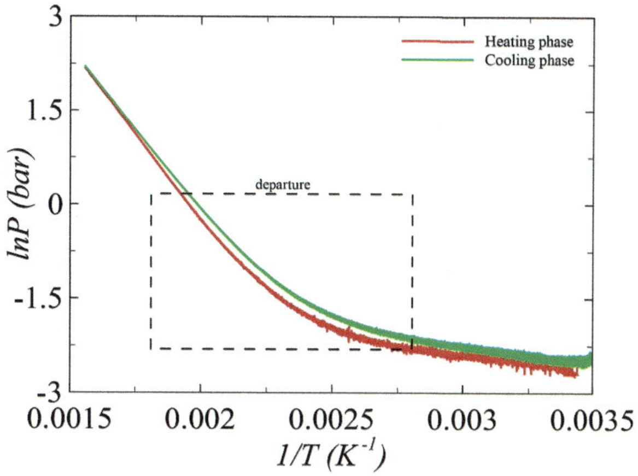

Fig. 7 Saturation curve of D_6 during the heating and cooling processes for $\Gamma = -0.045$ case

4 Conclusions and Future Works

Rarefaction shockwave experiments were conducted at different operating conditions using a high molecular complexity fluid (D_6). Wave speed measurements strongly show the existence of nonclassical effects when the fluid thermodynamic state falls in the negative Γ region (close to the critical point) in contrast to its ideal regime. The repeatability of the nature of wave propagation is also ensured by accurate measurements of the fluid pressure and temperature during the heating process. The studied conditions indicate possible thermal decomposition of the fluid, that needs to be further investigated and confirmed by GC analysis. Hence, a detailed investigation of the above-discussed experiments, along with the GC analysis, is an on-going research activity and will be presented in future studies. This set of experiments throws light on understanding nonclassical effects and the impact of the fluid thermal decomposition on the RSW experiments.

References

1. Guardone, A.L.B.E.R.T.O., Colonna, P., Pini, M., Spinelli, A.: Nonideal compressible fluid dynamics of dense vapors and supercritical fluids. Annu. Rev. Fluid Mech. **56**, 241–269 (2024)
2. Cramer, M.S., Best, L.M.: Steady, isentropic flows of dense gases. Phys. Fluids A **3**(1), 219–226 (1991)
3. Zamfirescu, C., Guardone, A.L.B.E.R.T.O., Colonna, P.: Admissibility region for rarefaction shock waves in dense gases. J. Fluid Mech. **599**, 363–381 (2008)
4. Colonna, P., Guardone, A.L.B.E.R.T.O.: Molecular interpretation of nonclassical gas dynamics of dense vapors under the van der waals model. Phys. Fluids **18**(5) (2006)
5. Fergason, S.H., Ho, T.L., Argrow, B.M., Emanuel, G.: Theory for producing a single-phase rarefaction shock wave in a shock tube. J. Fluid Mech. **445**, 37–54 (2001)
6. Colonna, P., Guardone, A.L.B.E.R.T.O., Nannan, N.R.: Siloxanes: a new class of candidate bethe-zel'dovich-thompson fluids. Phys. Fluids **19**(8) (2007)

7. Fergason, S., Guardone, A.L.B.E.R.T.O., Argrow, B.: Construction and validation of a dense gas shock tube. J. Thermophys. Heat Transf. **17**(3), 326–333 (2003)
8. Mathijssen, T., Gallo, M., Casati, E., Nannan, N.R., Zamfirescu, C., Guardone, A.L.B.E.R.T.O., Colonna, P.: The flexible asymmetric shock tube (fast): a ludwieg tube facility for wave propagation measurements in high-temperature vapours of organic fluids. Exp. Fluids **56**, 1–12 (2015)
9. Borisov, A.A., Borisov, A.A., Kutateladze, S.S., Nakoryakov, V.E.: Rarefaction shock wave near the critical liquid–vapour point. J. Fluid Mech. **126**, 59–73 (1983)
10. Guardone, A.L.B.E.R.T.O., Colonna, P., Casati, E., Rinaldi, E.: Non-classical gas dynamics of vapour mixtures. J. Fluid Mech. **741**, 681–701 (2014)
11. Chandrasekaran, N.B., Mercier, B., Colonna, P.: Nonlinear wave propagation in dense vapor of bethe–zel'dovich–thompson fluids subjected to temperature gradients. Phys. Fluids **33**(10) (2021)
12. Chandrasekaran, N., Michelis, T., Mercier, B., Colonna, P.: Preliminary experiments in high temperature vapours of organic fluids in the asymmetric shock tube for experiments on rarefaction waves (aster). In: International Seminar on Non-Ideal Compressible-Fluid Dynamics for Propulsion & Power, pp. 201–208. Springer (2022)
13. Chandrasekaran, N., Michelis, T., Mercier, B., Falsetti, C., Colonna, P.: Nonclassical wave propagation measurements in the high temperature vapor of D_6 with the asymmetric shock tube for experiments in rarefaction waves (aster). Exp. Fluids **65**(7), 1–16 (2024)
14. Calderazzi, L., Colonna di Paliano, P.: Thermal stability of r-134a, r-141b, r-13i1, r-7146, r-125 associated with stainless steel as a containing material. Int. J. Refrig **20**(6), 381–389 (1997)
15. Chandrasekaran, N.B.: Nonclassical gasdynamics: theory and experiments on nonlinear wave propagation in BZT fluids. Ph.D. Thesis, Delft University of Technology (2023)
16. P Colonna, N.R.N., Guardone, A.L.B.E.R.T.O.: Multiparameter equations of state for siloxanes:[(ch3) 3-si-o1/2] 2-[o-si-(ch3) 2] i= 1, · · · , 3, and [o-si-(ch3) 2] 6. Fluid Phase Equ. **263**(2), 115–130 (2008)
17. Pasetti, M., Invernizzi, C.M., Iora, P.: Thermal stability of working fluids for organic rankine cycles: An improved survey method and experimental results for cyclopentane, isopentane and n-butane. Appl. Thermal Eng. **73**(1), 764–774 (2014)

Towards Accurate and Repeatable Static Pressure Measurements in Expanding Organic Flows

Alessio Secchiaroli, Adam J. Head, Chiara Falsetti, Carlo De Servi, and Piero Colonna

Abstract This paper describes the operation of a pressure scanner, specifically the *Scanivalve DSA 3218* device, when used to characterize the pressure evolution along the expansion process of an organic vapor. The working fluid Hexamethyldisiloxane (MM) is expanded in the De Laval nozzle of the ORCHID facility for various inlet thermodynamic conditions: 220 °C and 4bar, 220 °C and 10.8bar, and 244 °C and 13.3bar. Pressure is measured along the nozzle walls using 16 pressure taps. Each of them is connected to a pressure line, which may be affected by varying rates of working fluid condensation, despite all pressure lines being purged with nitrogen before starting data acquisition. Fluid condensation increases the time response of the measurement system and potentially impacts the accuracy of the measurements. Repeated tests show that the steady-state values recorded by the pressure scanner once hydrostatic equilibrium is reached in the lines, are marginally affected by the amount of condensing fluid and the adopted purging strategy. This is observed for all pressure lines, including those connected to the taps placed before the nozzle throat, where substantial condensate accumulation occurs during testing as the thermodynamic state of the vapour is closer to the saturation curve. The experimental campaign documented in this paper paves the way for further refinement of the pres-

A. Secchiaroli (✉) · A. J. Head · C. Falsetti · C. De Servi · P. Colonna
Propulsion and Power, Delft University of Technology,
Delft, The Netherlands
e-mail: a.s.secchiaroli@tudelft.nl

A. J. Head
e-mail: a.j.head@tudelft.nl

C. Falsetti
e-mail: c.falsetti@tudelft.nl

C. De Servi
e-mail: c.m.deservi@tudelft.nl; carlo.deservi@vito.be

P. Colonna
e-mail: p.colonna@tudelft.nl

C. De Servi
Energy Technology Unit, VITO, Mol, Belgium

A. Giauque et al. (eds.), *Proceedings of the 5th International Seminar on Non-Ideal Compressible Fluid Dynamics for Propulsion and Power*, ERCOFTAC Series 32,
https://doi.org/10.1007/978-3-031-92695-2_17

sure scanner measurement procedure for non-ideal vapour flows and its application to the characterization of expansion processes in small power capacity ORC turbines.

Keywords Static pressure measurement · Expanding non-ideal vapour flow · Pressure scanner · Condensation effects

1 Introduction

Common experimental techniques can be applied to non-ideal compressible fluid dynamics (NICFD) flows, but customized procedures are needed as experiments with organic vapors entail technical challenges such as vapor condensation, fluid thermal decomposition, uncertainty in the thermo-physical properties, etc. Different examples of these adaptations are documented in [1–5]. In this study, a multichannel pressure scanner developed by *Scanivalve* is used to measure the static pressure distribution along the walls of a planar de Laval nozzle. Static pressure measurements, paired with velocity measurements, allow for the characterization of pressure losses in expanding organic flows, provided that no work extraction takes place, thermal losses are minimal, and an accurate equation of state for the fluid of interest is available. This information is crucial for validating codes used in the design of turbomachines operating with organic fluids. An attempt to enable the measurement of total pressure in organic supersonic flows was presented in [6], where the authors documented the development of a pneumatic system for calibrating pressure probes for NICFD flows. Key differences with respect to the *Scanivalve* setup used in this work include the use of a dedicated transducer for each pneumatic line, and no purging system to prevent the condensate from reaching the transducers. Conversely, the *Scanivalve* scanner allows for the connection of multiple pressure taps to the same sensor. However, its transducers may be damaged if they come into contact with condensate accumulating in the pneumatic lines. Additionally, the pressure measurement system described in [6] is designed for transient operation, while the *Scanivalve* system is intended for steady-state experiments and long testing times.

The aim of this research is to assess how the *Scanivalve* measurement system operates under various NICFD flow conditions and devise a strategy to mitigate vapor suction in the pressure lines and increase the repeatability and duration of the experiments.

2 Experimental Setup

The ORCHID setup is a fully instrumented ORC facility featuring two test sections: a de Laval nozzle with optical access and a turbomachinery test bench. This setup allows for steady-state experiments in NICFD flows over a wide spectrum of thermodynamic conditions. The thermal power required to evaporate the working fluid is supplied by a 400 kW electric oil heater with a maximum oil temperature of 400 °C. The cooling power for condensation is provided by a water loop with an integrated air cooler. A detailed description of the facility is reported in [7]. The working fluid employed in the experimental campaign is hexamethyldisiloxane, hereafter indicated as MM. Moreover, the test section of interest for this work is the de Laval nozzle, which includes a settling chamber upstream of the supersonic nozzle for measuring fluid total conditions and a receiver to maintain a constant back pressure for the expansion process. The settling chamber features two wire meshes and a honeycomb screen to straighten the flow before it reaches the nozzle. This component features a 2D profile, which has been designed with the Method of Characteristic (MoC) for total inlet conditions of 252 °C and 18.4bar(a). The outlet conditions are set to achieve a supersonic flow characterized by a Mach number over 2 at the nozzle exit. More details on the design methodology can be found in [8].

3 Method

The pressure scanner, manufactured by *Scanivalve*, features 16 acquisition ports connected to pressure taps manufactured along the nozzle profile, whose positions are indicated in Table 1, as well as 2 taps in the flanges of the settling chamber to

Table 1 Pressure taps position and labels

Tap number	Axial coordinate	Scanivalve
	/mm	Port Number
T002	n.a	PSV001
$T003_s$	15.96	PSV002
T003	15.96	PSV003
T004	24.72	PSV004
T005	33.55	PSV005
T006	42.4	PSV006
$T006_s$	42.4	PSV007
T008	47.44	PSV008
T009	49.9	PSV009
T010	52.57	PSV010
T011	55.01	PSV011
T012	57.5	PSV012
T013	59.91	PSV013
T014	62.53	PSV014
T017	91.3	PSV015
$T017_s$	91.3	PSV016

The second tap is located on the flange connecting the settling chamber and test section

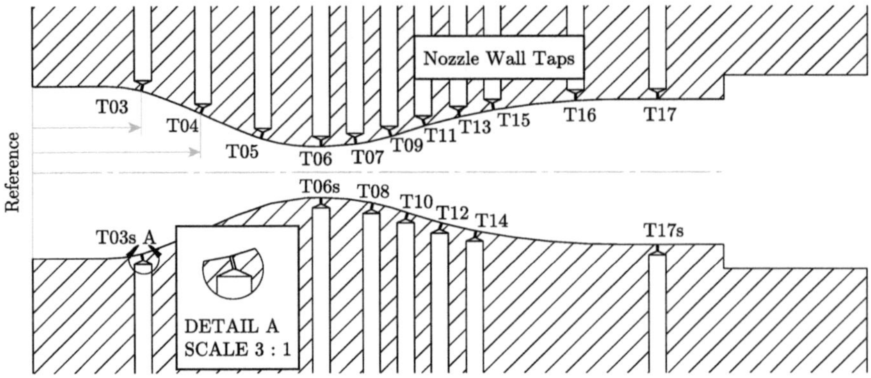

Fig. 1 Nozzle cross-section showing the pressure taps positions

monitor the pressure drop across that component. Three pairs of pressure taps are positioned at the same axial location but on two opposite sides of the nozzle profile to verify flow symmetry, see Fig. 1. The *Scanivalve* features 3 pressure sensors with different full scale. Their accuracy is $\pm 0.05\%$ of the full scale. The pressure ports can be grouped based on the full-scale of the sensor to which they are connected, namely: PSV001 - 004 up to 34.5bar(g), PSV005 - 010 up to 17.2bar(g), and PSV011 - 016 up to 6.8bar(g). The pressure sensors used to measure the total pressure in the settling chamber and the nozzle back pressure in the receiver, labeled respectively as PT011 and PT004, are a *WIKA UPT 20* piezoelectric transducers with an expanded uncertainty of $\pm 0.1 bar$.

Figure 2a displays the nozzle test section with the *Scanivalve* box in the top left corner. Each pressure tap line features a liquid trap, as shown in Fig. 2b, to collect the condensate by gravity. The liquid level can be monitored through a sight glass. A ball valve installed on top of each trap connects to the test section and can be used to isolate the line from the process when necessary, while a bottom valve connects via a collecting pipe to a tank. The bottom valve is used to discharge condensate accumulated during testing in a controlled way, thus allowing for a longer duration and easier repetition of the experiments. To prevent air from entering the pressure lines while emptying the liquid traps, these are flushed by injecting nitrogen at 3bar(a). The red caps in Fig. 2b serve as a safety mechanism to prevent accidental releases of condensate into the atmosphere.

The *Scanivalve DSA* 3218 system has been employed in different applications, but not for pressure measurements in NICFD flows. To the authors' knowledge, the ORCHID is the only facility to utilize a pressure scanner for experiments involving organic vapours. For this reason, an ad-hoc operating procedure was devised to ensure the safe and reliable operation of the *Scanivalve*. The main steps of the operation procedure, which is partly automated by means of a *LabView* program, can be summarized as follows:

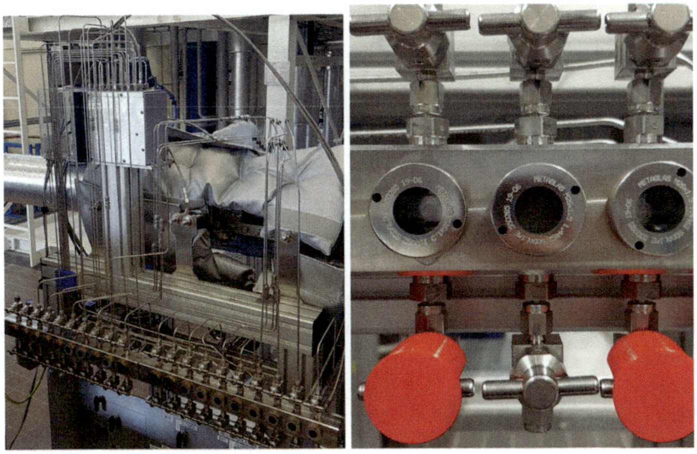

(a) (b)

Fig. 2 **a** Nozzle housing and the *Scanivalve* measurement system. **b** Close-up of three liquid traps

- Once the targeted nozzle inlet temperature and pressure (T_0, p_0) are reached at the outlet of the evaporator, the test section is opened, and the operating conditions are left to stabilize for a few minutes to ensure steady-state flow in the nozzle test section.
- The pressure lines are purged with nitrogen at a slightly higher pressure than the one at the nozzle inlet to remove possible condensate accumulated before the test. However, the back-pressure of the nozzle increases due to the build-up of nitrogen in the condenser, as shown in Fig. 3. The purging of the lines is needed to prevent condensate from reaching the pressure scanner sensors and eventually damage them during data acquisition, as well as to avoid the formation of a liquid head in the lines that can affect the pressure measurements.
- After the completion of the purging, the nitrogen in the lines should be in mechanical equilibrium with the expanding flow in the nozzle. This implies that the pressure in each line should be equal to the pressure of the flow at the corresponding pressure tap.
- Once the operating conditions of the setup reach steady-state again, the pressure measurements are recorded. A 'Zero Offset Calibration' is carried out at the first data acquisition to correct for the piezoresistive sensors' drift caused by variations in ambient temperature.

For more details about the operation of the *Scanivalve* system, the interested reader is referred to Appendix C in Ref. [7].

Three experimental conditions were tested and analyzed, SP1, SP2, and SP3, which are reported in Table 2. The Compressibility Factor (Z_0) of MM at the nozzle inlet and the representation of the expansion process of each test run in the $T - s$ diagram of Fig. 4 are obtained using the software REFPROP from NIST [9].

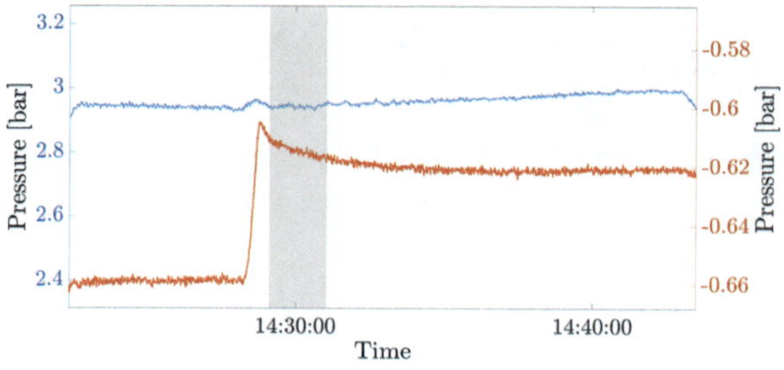

Fig. 3 Recording of pressure upstream and downstream of the nozzle during the experiment with the pressure scanner. The experimental conditions are those of SP1. (– PT011, – PT004, – Data acquisition window)

Table 2 Experimental conditions

Case	Total Pressure $P_0/bar(g)$	Total Temperature T_0°C	Compressibility Factor Z_0	Back-pressure (PT004) $P_b/bar(g)^{@}$
SP1	4	220	0.9150	−0.612
SP2	10.8	220	0.7106	−0.138
SP3	13.3	244	0.7125	0.101

As measured after pressure scanner purging and stabilization of the condensing pressure

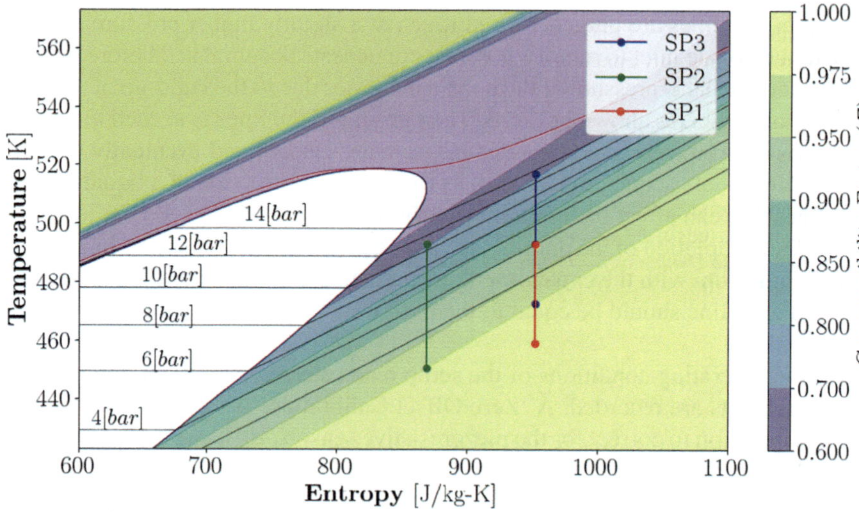

Fig. 4 Representation of the expansion processes associated with the three experimental conditions in the $T - s$ diagram

The experimental conditions chosen for SP1 match those of a previous Particle Image Velocimetry (PIV) experiment [5], while the SP2 and SP3 conditions are selected based on the findings in [10]. Based on numerical studies, the authors of this work concluded that the expansion processes realized in SP2 and SP3 can be accurately modeled by using the Peng-Robinson Equation of State (EoS) [11]. The aim of the experiments outlined in this paper is then to provide a dataset of accurate measurements to validate, in the future, that hypothesis.

4 Results

Figure 5 shows the pressure evolution in the lines connected to the taps placed upstream the nozzle throat, namely from PSV001 to PSV007, as recorded by the pressure scanner during the tests. During the three experiments (SP1-2–3), a transient response in the lines was observed, except for line PSV001. In contrast, no transient seemingly occurred in the lines of the pressure taps placed in the diverging section of the nozzle in Fig. 5, for the sake of clarity, the pressure measurements pertaining to these lines have not been displayed. These trends can be explained by the following considerations:

- The purging of the pressure lines is performed following the procedure indicated by the manufacturer. The assumption that the pressure in the lines after the purging would be equal to that of the flow in correspondence with the specific pressure tap proved to be false. Clear evidence thereof has been obtained by isolating each line from the process before completing the purging procedure and then starting data acquisition. The measurements revealed that pressure was at a similar value in all the lines. The recorded pressure closely matched the pressure maintained in the sensor housings of the pressure scanner to protect the sensors from an overpressure condition during line purging.
- If the pressure lines are not adequately pressurized with nitrogen through the purging procedure, significant quantities of working fluid enter and condense, since the temperature in the lines is close to the ambient one. The condensation causes a delay in pressure buildup in the lines, leading to the drawing of more fluid from the nozzle. As a result, the condition of hydrostatic equilibrium is only achieved when the displacement caused by the condensate accumulated in the lines is sufficient to pressurize the nitrogen up to the flow pressure at the tap location. This occurs after a transient response which can last several minutes. Its duration depends on the amount of nitrogen remaining in each line after purging, as well as the possibility of flow choking in the pressure taps, which represent the point of minimum cross-sectional area for the vapor flow into the pressure lines. Such a situation tends to occur in all the lines connected to the taps placed up to the nozzle throat and is analogous to the mass sink effect described in [6]. An exception is the response of the line connected to PSV001, as discussed more in detail later.

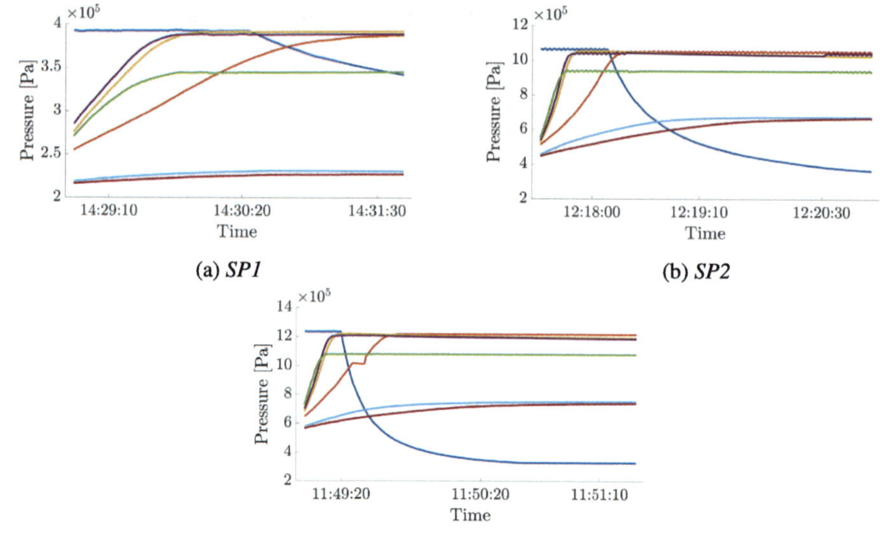

(a) *SP1*

(b) *SP2*

(c) *SP3*

Fig. 5 Pressure recorded by the *Scanivalve* during the tests. (– PSV001, – PSV002, – PSV003, – PSV004, – PSV005, – PSV006, – PSV007)

- If the pressure in the lines after the purging is higher than that in the flow, the excess nitrogen is discharged through the pressure taps. This transient was not observed during the tests, because it lasted for a shorter time than the interval between the end of the purging procedure and the start of data recording. This situation may have occurred in all the pressure lines connected to the taps placed downstream of the nozzle throat.

As mentioned above, the data recorded by the pressure scanner for the line connected to PSV001 does not show any transient response after the start of the acquisition, despite the significant condensation rate. The amount of condensate was sufficient to rapidly fill up the liquid trap, to the point that the line had to be isolated from the nozzle by closing the valve installed upstream of the liquid trap itself. The moment the valve was closed can be recognized in Fig. 5, as it corresponds to the start of the pressure decrease in the line. Such a drop in pressure is due to further condensation of the working fluid in the line occurring after the valve is closed. The hypothesis put forward to explain the absence of any appreciable transient response in the line of PSV001 at the beginning of the data acquisition period is that the diameter of that pressure tap is much larger than in the original design specifications. A numerical model will be implemented to assess the validity of this hypothesis.

If the liquid traps become flooded, the condensate accumulation in the pressure lines arguably leads to a measurement error due to potential differences at the end of the pressure transient in the liquid level in the two lines at the top of the liquid trap. This measurement error is directly proportional to the difference in liquid elevation in the two lines with the proportionality constant given by the specific weight of the fluid. However, the systematic error in the pressure measurement is limited and it is expected to be at most similar to the sensors' uncertainty. In a worst-case and unrealistic scenario, where the difference in liquid level in the two lines reaches a maximum of 80 cm, the resulting systematic error in the pressure measurements is estimated to be approximately 6000Pa. This compares to an expanded uncertainty of the sensor connected to the taps T002 - T004 of 1775Pa, and 910Pa for the sensor of the taps T005 - T006s.

To prove that the measurement error does not compromise the reliability of the devised measurement chain, the experiment performed for the SP3 conditions was repeated eight times testing different purging strategies. These differed in the order in which the various lines were opened during purging, the isolation of the lines from the process before closing the nitrogen supply line, and the pre-emptying of the liquid traps. Such operation was performed by opening the valve placed at the bottom of the liquid traps with the simultaneous injection of nitrogen in the lines. The tests showed that the purging strategy has an impact on the amount of nitrogen that is injected into the working fluid loop and the duration of the pressure transient in the lines connected to the taps placed from the nozzle inlet up to the throat. At the same time, the recorded steady-state pressure values were consistent in all tests, as shown in Table 3 which reports the average and the standard deviation of the final pressure established in the lines. The standard deviation is small regardless of

Table 3 Absolute uncertainty of *Scanivalve* sensors U (coverage factor k = 1), and average and standard deviation of the steady state pressure measured by the pressure scanner for the eight tests run for the experimental conditions of SP3

Pressure line	Mean [bar(a)]	σ [bar(a)]	U [bar(a)]
PSV001	13.1121	0.0096	0.0089
PSV002	12.9758	0.0214	0.0089
PSV003	12.9976	0.0156	0.0089
PSV004	12.8365	0.0184	0.0089
PSV005	11.5967	0.0136	0.0046
PSV006	8.2330	0.0676	0.0046
PSV007	8.2724	0.0430	0.0046
PSV008	4.4918	0.0041	0.0046
PSV009	3.4793	0.0082	0.0046
PSV010	2.7942	0.0043	0.0046
PSV011	2.6221	0.0097	0.0020
PSV012	2.4237	0.0031	0.0020
PSV013	2.4161	0.0070	0.0020
PSV014	2.2547	0.0037	0.0020
PSV015	1.4479	0.0077	0.0020
PSV016	1.4547	0.0152	0.0020

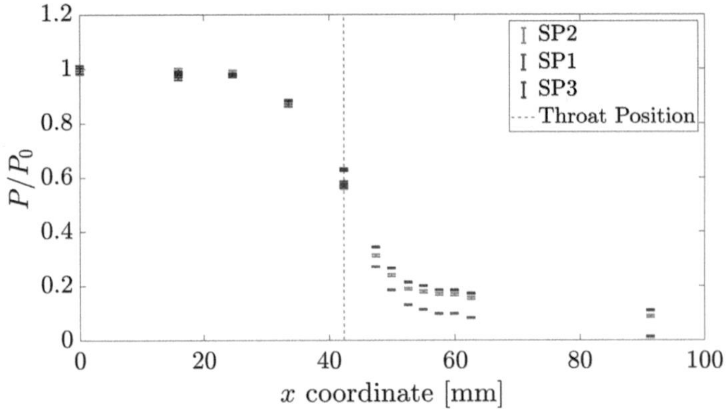

Fig. 6 Pressure along the nozzle measured for the three experimental conditions. The pressure profiles are non-dimensionalized against the total inlet pressure

the tap location, though tends to be larger than the uncertainty associated to sensor accuracy. This can be attributed to variations in the nozzle operating conditions, as well as the inherent variability in test execution and data post-processing, and only to a minor extent to differences in the amount of condensate accumulated in the lines. The relative measurement uncertainty is, indeed, comparatively higher for the pressure lines unaffected by fluid condensation, with the exception of the pressure measured in correspondence to the nozzle throat. If fluid condensation does affect measurement quality, it may do so, though marginally, only for the pressure lines connected to the taps at the nozzle throat.

Figure 6 shows the pressure profile along the nozzle reconstructed via the pressure scanner for the three operating conditions investigated in this work. The values are non-dimensionalized to the respective inlet nozzle pressure for comparison purposes. It is noticeable that the pressure profiles for the SP2 and SP3 conditions are close to each other, given the similar pressure ratio and evolution of the compressibility factor along the expansion.

5 Conclusions

This work analyses the operation of a pressure scanner, specifically the *Scanivalve DSA 3218* system, when used to characterize a supersonic flow of an organic vapor. The experiments involved three expansion processes of the organic fluid MM in the de Laval nozzle of the ORCHID facility, starting from thermodynamic states of different levels of vapour non-ideality. The tests show that:

1. The purging procedure devised to protect the pressure scanner sensors proved unsuccessful in preventing the working fluid from entering the pressure lines connected to the taps placed before and at the nozzle throat during acquisition.

This does not pose a risk to the integrity of the sensors, as the working fluid condenses and is accumulated in the liquid traps installed at the lowest point of the pneumatic lines.
2. The pressure at the tap location is measured by the pressure scanner when the fluid in the lines, whether it is nitrogen only or a mixture of MM and nitrogen, is in hydrostatic equilibrium with the flow in the nozzle. If the working fluid enters the lines after the purging procedure is completed, reaching pressure equilibrium is delayed due to vapour condensation.
3. The steady-state pressure recorded by the *Scanivalve* once hydrostatic equilibrium is reached is affected by the presence of condensate in the lines only if the liquid traps become flooded. However, the corresponding error in the pressure measurement is marginal. This has been proven by repeating the same experiment while implementing various purging strategies. The steady-state pressures measured in these different tests fell within the expected uncertainty range for these types of measurements, thus proving the reliability of the developed measurement chain in characterizing the pressure of the expanding flow in the ORCHID nozzle.

To facilitate the purging of the pressure lines and enable longer acquisition times, the liquid traps are connected via a discharge line to a tank such that the condensate can be removed during the tests. This discharge line may also be exploited to minimize the amount of nitrogen that is injected into the loop during the purging of the pressure lines, which causes a non-negligible increase in the condensing pressure of the setup. Further efforts will also be dedicated to improving the purging strategy in order to minimize the quantity of working fluid entering the pressure lines. Preliminary tests have shown that through simple modifications to the existing purging sequence, it is feasible to prevent vapor from entering the lines during experiments with a nozzle inlet pressure of up to 10bar(a). Moreover, the devised pressure measurement system will be exploited in the future to characterize the expansion process in the stator of a small power capacity ORC turbine currently being installed in the ORCHID setup.

References

1. Spinelli, A., Cammi, G., Zocca, M., Gallarini, S., Cozzi, F., Gaetani, P., Dossena, V., Guardone, A.: Experimental observation of non-ideal expanding flows of siloxane MDM vapor for ORC applications. Energy Proc. **129**, 1125–1132 (2017)
2. Gallarini, S., Cozzi, F., Spinelli, A., Guardone, A.: Direct velocity measurements in high-temperature non-ideal vapor flows. Exp. Fluids **62**(10), 199 (2021)
3. Sundermeier, S., Matar, C., Cinnella, P., aus der Wiesche, S., Hake, L., Gloerfelt, X.: Experimental and numerical study of transonic flow of an organic vapor past a circular cylinder. In: Proceedings of the 4th International Seminar on Non-Ideal Compressible Fluid Dynamics for Propulsion and Power, pp. 209–216, London, United Kingdom. Springer (2023)
4. Wiesche, S.A.D.: Experimental investigation techniques for non-ideal compressible fluid dynamics. Int. J. Turbomach. Propul. Power **8**(2), 11 (2023)
5. Michelis, T., Head, A.J., Majer, M., Colonna, P., De Servi, C.: Assessment of particle image velocimetry applied to high-speed organic vapor flows. Exp. Fluids **65**(6), 90 (2024)

6. Conti, C.C., Fusetti, A., Spinelli, A., Gaetani, P., Guardone, A.: Pneumatic system for pressure probe measurements in transient flows of non-ideal vapors subject to line condensation. Measurement **192**, 110802 (2022)
7. Head, A.J.: Novel experiments for the investigation of non-ideal compressible fluid dynamics. Ph.D. thesis, Delft University of Technology, Delft, the Netherlands (2021)
8. Anand, N.: Progress in CFD-based automated design with application to unconventional turbomachines. Ph.D. thesis, Delft University of Technology, Delft, the Netherlands (2021)
9. Lemmon, E.W., Bell, I.H., Huber, M.L., McLinden, M.O.: NIST Standard Reference Database 23: Reference Fluid Thermodynamic and Transport Properties-REFPROP, Version 10.0, National Institute of Standards and Technology (2018)
10. Fernandes, D.S.C., Majer, M., Head., A.J.: Towards a thermodynamic validation map for non-ideal compressible flows. In: Submitted to 5th International Seminar on Non-Ideal Compressible-Fluid Dynamics for Propulsion & Power, Lione, France (2024)
11. Smith, J.M., Van Ness, H.C., Abbott, M.M., Swihart, M.T.: Introduction to Chemical Engineering Thermodynamics. McGraw-Hill Education, New York, USA (2022)

Commissioning and Preliminary Testing of the NextORC Facility

Omnia El-Baz, Tala El Samad, Mohsen Ghavami, Martin T. White, and Abdulnaser Sayma

Abstract Turboexpanders are recommended for medium- and low-temperature organic Rankine cycle applications as they can handle large enthalpy drops thus achieving high pressure ratio across a single stage with high efficiency but they have a low degree of reaction where most of the enthalpy rise occurs in the stator section. The requirement for compactness and operational flexibility intersects with challenges posed by a low compressibility factor, and non-ideal gas effects. Several numerical models have been developed to investigate the non-ideal effects that occur, however, experimental data is still insufficient. Within this paper, progress in the development of the NextORC experimental facility, which includes a converging-diverging nozzle representative of a nozzle guide vane within a turbo-expander, is presented. The test facility is designed to provide an enhanced understanding of supersonic flow dynamics through the nozzle with outlet Mach number of 2. This paper outlines the setup, functionality, and initial experiments conducted at the test facility which is designed to operate at pressures up to 15 bar, temperatures reaching 120 °C, and compressibility factor down to 0.75. The results from initial commissioning tests are detailed, which demonstrate that steady-state conditions can be achieved for seven different inlet total conditions with pressures from 7 to 13 bar. Although supersonic flows were not established, the subsonic flow pressure distributions were recorded during the commissioning phase.

O. El-Baz (✉) · T. El Samad · M. Ghavami
City St George's, University of London, London, UK
e-mail: Omnia-Ahmed.Elbaz@citystgeorges.ac.uk

T. El Samad
e-mail: T.ElSamad@citystgeorges.ac.uk

M. Ghavami
e-mail: M.Ghavami@citystgeorges.ac.uk

M. T. White
University of Sussex, Brighton, UK
e-mail: martin.white@sussex.ac.uk

A. Sayma
Brunel University London, London, UK
e-mail: abdulnaser.sayma@brunel.ac.uk

A. Giauque et al. (eds.), *Proceedings of the 5th International Seminar on Non-Ideal Compressible Fluid Dynamics for Propulsion and Power*, ERCOFTAC Series 32,
https://doi.org/10.1007/978-3-031-92695-2_18

Keywords Organic rankine cycle · Expander · Converging-diverging nozzle

1 Introduction

Organic Rankine cycles (ORCs) are one of the most suitable solutions for clean energy production from low-to-medium temperature ($100 - 400°C$) heat sources and could have a positive contribution in achieving net zero targets for both power generation and industrial applications. ORC technology can be used to recover waste energy produced from industrial processes and renewable energy sources such as solar, biomass, and geothermal applications [1]. The stator vanes found within ORC turbines commonly operate within the domain of non-ideal compressible flow dynamics (NICFD), which occurs when the gas dynamics of fluids do not obey the ideal gas law. This non-ideal thermodynamic region includes single-phase states at thermochemical equilibrium resulting in the fluid being either a dense vapour, a supercritical fluid, or a compressible liquid. Specifically, in the case of vapours under subcritical pressure conditions, the compressibility factor (Z) is typically smaller than 1, which categorizes it as a dense vapour [2]. The high molecular complexity of ORC working fluids with a low speed of sound can lead to supersonic and transonic conditions at the exit of the turbine rotor, which affect turbine loss mechanisms, including trailing edge and shock losses. Additionally, as molecular complexity increases, non-ideal gas dynamic effects become more pronounced; this can result in quantitative and qualitative differences.

Another thermodynamic quantity defining the real gas effect is the fundamental derivative of gas dynamics (Γ). The propagation of waves through dense vapour depends on the variation of density and the speed of sound. For the region where $0 < \Gamma < 1$, the pressure and speed of sound relationship can be inverted from the ideal gas behaviour [3]. It means that the speed of sound increases during isentropic expansion and decreases during compression which causes the challenge to appear for ORC turbine design.

The performance estimated by numerical studies and computational fluid dynamics simulations requires suitable experimental data for validation and verification. While efforts are in progress, most non-ideal gas dynamic effects still lack experimental verification [2]. Recently, there has been progress in experimental research with the development and running of test facilities to gain understanding of NICFD behaviour for ORC applications. The design of these facilities varies between blowdown and continuous operations. For instance, The Test Rig for Organic Vapors (TROVA) in Politecnico di Milano is a blow-down wind tunnel designed to investigate the characterization of organic vapour flow in the expansion process of ORC turbines [4]. The expansions of siloxanes, such as MM and MDM, are studied using a double passage type Schlieren measurement system. One of the continuous working test facilities is the closed Loop Organic Wind Tunnel (CLOWT) in Münster University of Applied Sciences which aims at obtaining loss correlations for performance predictions of ORC expanders [5, 6]. The expansion of Novec 649 through a cascade test section and a test bench for an ORC expander is conducted, with pressure measurements taken for flow characterization. The ORCHID facility (Organic Rankine Cycle Hybrid Integrated Device) which is a Rankine wind tunnel (continuously working) is designed and operated at TU Delft to enhance fundamental studies on NICFD flows and test ORC components, e.g., ORC turbines and heat exchangers [7]. The expansion of siloxane MM was performed through a converging-diverging nozzle (CD) with pressure measurements and schlieren visualization. Additionally, a blowdown facility was designed and built at Imperial College London for the investigation of small-scale, high Mach number flow physics and the validation of computational methods [8]. A test section with a CD was used to study the expansion of R1233zd(E) with pressure measurement of the flow. Although these facilities have enabled significant steps in the field of NICFD, the available experimental data is still in the development phase, and contributing with additional data is necessary to enhance the existing dataset which is the primary motive behind the building and commissioning of NextORC test rig at City St George's, University of London.

The NextORC test facility is a dual-purpose facility that aims at improving the fundamental understanding of the performance of turbo- and twin-screw expanders for small-scale ORC systems (100 to 150 kW), utilizing low-temperature heat sources. The rig is designed to be capable of investigating the expansion of both superheated vapour and two-phase flows. The facility is of continuous operation driven by a twin-screw refrigeration compressor which removes the need for a large heat source. The continuous operation enables long measurement times and the possibility of reaching steady-state. Within the test rig, a cascade with a converging-diverging nozzle capable of reaching outlet Mach number of 2 has been designed and installed to replicate the expansion of the flow through the nozzle portion of a stator nozzle guide vane of a turboexpander operating with superheated flow in the dense vapour region. Typically, small-scale ORCs employ a single-stage turbine characterized by a high-pressure ratio, resulting in high-speed flows at the exit of the stator nozzle, with Mach numbers exceeding 2. This supersonic flow of organic fluid constitutes a critical aspect of turbine design, underscoring the necessity to comprehend the fundamentals of flow phenomena under such conditions. This paper begins with a detailed description of the components, measurement equipment, and estimation of uncertainty. This is followed by a discussion of the operation of the test rig, and then reports the initial results and analysis of the commissioning tests. The data sets generated include static pressure measurements along the nozzle centreline. Finally, the paper concludes with an explanation of future work to come.

2 NextORC Experimental Setup

2.1 Main Components

Figure 1 (left) depicts a schematic diagram of the facility which shows the main components for the two working loops of the NextORC test rig. This section contains the details and specifications of the major components

Compressor: The facility is driven by an SRC-S series twin screw oil-injected compressor(SRC-S series) with a nominal power of 45 kW, a pressure range of 19–25 bar, and capable of delivering a flow rate of 175 m^3/h. The compressor is used to circulate the working fluid and reach the target superheated vapour conditions required at the inlet of the cascade. After passing through the compressor, the flow proceeds into the cascade and then exits the cascade before entering the condenser.

The cascade test section: The test section is a custom-build design that consists of six components/processes. First, the high-pressure and temperature vapour leaving the compressor enters an upstream diffuser and settling chamber, which aims to slow down the flow and reduce disturbances caused by the compressor, valves, or during upstream turning of the flow. A contraction zone is placed after the settling chamber to accelerate the flow, while preventing flow separation and ensuring a high flow uniformity. After the contraction zone, the flow is delivered to the converging-diverging nozzle inlet. The working fluid expands in the converging-diverging nozzle to reach the required Mach number at the outlet of the test section. The test section is a rectangular steel channel within which different geometries can be placed. Figure 1 (right) shows the whole facility, including the test section components and nozzle assembly respectively. Full details of the design of the cascade have been published in a previous study [9].

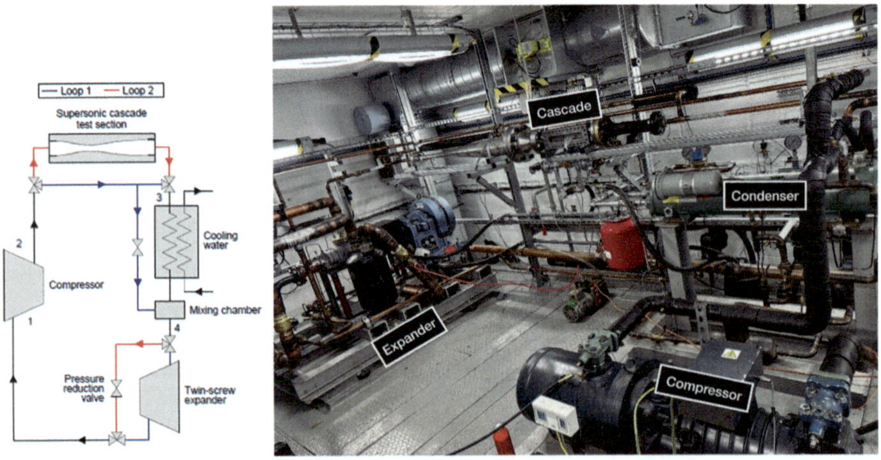

Fig. 1 (left): Schematic diagram (right): photo of NextORC test facility

Condenser: The flow is required to return to a partial-liquid state after the expansion process through the cascade, or before entering the expander in case of two-phase expansion studies, which is done using a condenser. A K573H BITZER shell and tube water-cooled condenser with finned tubes is used to condense the working fluid. The maximum working pressure of the condenser is 33 bar The lab process water feed is connected to the cold side of the condenser, and a pressure regulator is connected to the water side of the condenser to maintain a steady condensing pressure. The lab feed water is cooled by a pressurised closed loop cooling cycle using a shell and tube heat exchanger that uses chilled water from the building at 10 °C. The regulator opens when the condenser side (inlet side) pressure reaches the predetermined level.

Mixing chamber: the mixing chamber is installed to enable the study of two-phase expansion with different inlet vapour qualities. It is constructed from a small copper cylinder where part of flow leaving the compressor (expander tests) or cascade test section (cascade tests) is mixed with the flow from the condenser, controlled with manual valves, to reach the required vapor quality.

Expander and pressure reduction valve: After the condenser, the flow either passes through the expander, or through a pressure reduction valve, during which the pressure is reduced to the compressor suction pressure. Although it was originally envisioned that the expander would not be used when running cascade tests, the expander can be utilised to provide further control over the cascade conditions when operating the cascade.

Working fluid: R245fa was selected for the NextORC test facility experiments because of its availability, thermal stability, non-toxicity, non-flammability, and suitability for low-temperature applications. The thermophysical properties of R245fa are mentioned in Table 1. While R245fa is popular in current ORCs, there are uncertainties for the future use of this working fluid mainly because of the general phase-out of hydrofluorocarbons (HFCs) due to their increased global warming potential, thus alternatives are being researched [1]. Despite this, R245fa was still chosen for this test facility due to compressor limitations.

2.2 *Instrumentation and Measurement Equipment*

Pressure measurement: Pressure transducers are connected to static pressure taps, which were machined with a diameter of 0.8 mm in the side wall of the nozzle

Table 1 Thermophysical properties of R245fa

		Normal boiling point		
Fluid	Molecular mass (g/mol)	°C	T_C(°C)	P_C (MPa)
R245fa	134.05	15.1	154.0	3.65

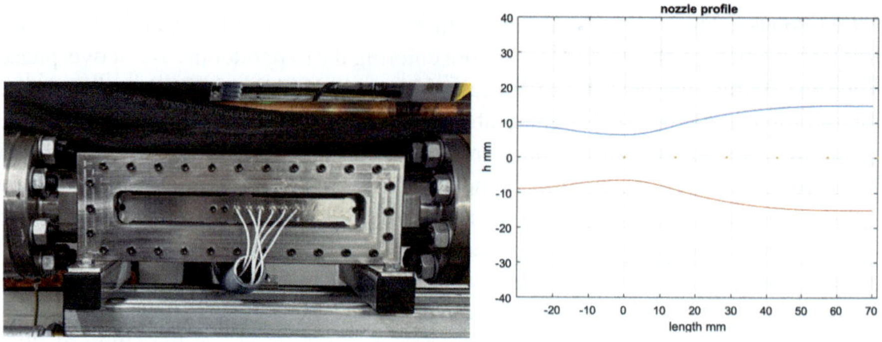

Fig. 2 Left: photo of the nozzle casing. Right: Sketch for nozzle profile showing pressure transducer location

chamber. Six Kulite XTEL-190SM pressure transducers are employed for measuring static pressure along the centreline of the converging-diverging nozzle. These transducers have a range of 0–17 bar and an output of 0–100 mV, which is amplified and converted to current separately. The combined uncertainty for non-linearity, hysteresis, and repeatability is $\pm 0.1\%$ of the full-scale output and the selected transducers are designed to withstand high temperatures and frequencies. The transducers are attached to the side of the cascade as shown in Fig. 2. The alignment of data between electrical and physical values is established via a calibration procedure in LabVIEW. For initial calibration, the process involves applying no pressure to the sensors and adjusting the zero setting to ensure a zero-pressure reading. Then pressure with known values was applied using a calibrated pressure kit. Pressure values are corrected on LabVIEW using the linear scaling method. First 4 mA was set at hard vacuum at 20 °C, then 20 mA span was set on amplifier at 1.7 MPa absolute using dry N_2 charge at 20 °C. Fine-trim the zero added in the software, per each transducer, to set as close as possible to 0 display at hard vacuum, 20 °C. Interim points are then taken on declining dry N_2 pressure, as the test charge is purged out, to check linearity. Finally, the pressure was put down to a hard vacuum to ensure no drift at zero. No external temperature compensation has been applied to the system, as the pressure transducer operates within its specified temperature range. According to the datasheet, the built-in compensation mechanisms account for thermal effects, such as thermal zero shift and sensitivity shift, within the operating temperature range of −65°F to +525 °F (cascade transducers XTEL-190.). The transducer's performance, including any thermal shifts, remains within acceptable limits, as outlined in the manufacturer's specifications. Pressures throughout the rig, excluding the cascade, are measured with Honeywell, PX3AF1BH020BSAAX transducers which have a Total Error Band (TEB) of $\pm 1.0\%$ Full Scale Span (FSS) from -20 °C to +85 °C and the expanded uncertainty of $\pm 2.0\%$ of FSS assuming 95% confidence level.

Temperature measurement: The temperature is measured throughout the rig, including at the cascade inlet and outlet, before and after compressor and expander,

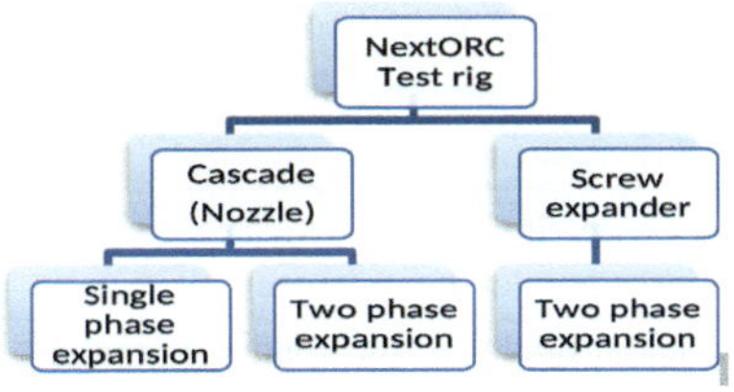

Fig. 3 Operation modes for the test rig

as well as at the liquid line of the condenser. Temperature is measured using PT100 resistance temperature platinum sensors. There are several ports on the rig where temperature sensors are inserted at the centreline of the flow line, including the inlet and outlet ports of the compressor, the screw expander, and the cascade inlet. Wall-mounted sensors are used in the rest of the rig, and the temperature difference between the measurement point and the refrigerant is determined by considering only the heat transfer resistance of the wall. This is because these temperature sensors are insulated from the surroundings to minimise parasitic heat transfer. These sensors have a temperature range of -75 °C to +250 °C with expanded uncertainty ranging between $\pm$ 0.6 and $\pm$ 3. Although thermocouples have a faster response time, resistance temperature detectors (RTDs) were chosen for their higher accuracy and stability, as the tests rig is designed to study flow under steady-state conditions.

3 Commissioning and Operation

3.1 Operational Procedure

The NextORC test rig can be operated in three different modes depending on the intended study. As shown in Fig. 3, the test rig can be tested for the single-phase and two-phase flows through the nozzle, or for the performance testing of the screw expander which will be taking place only in two-phase flow conditions, in line with the project's objectives to enhance the fundamental understanding of screw expander performance during two-phase expansion. This is in contrast to single-phase flow expansion, which has been extensively studied in screw expanders. For the case of the present paper, the test rig is operated within single-phase vapour flow through the nozzle.

Operation: The compressed vapour in superheat state (hot gas) leaves the compressor discharge and enters the cascade assembly first diffuser with of three mesh screens then the conditioning chamber where it passes through a honeycomb that works as a flow straightener and another turbulence breaker screen. The flow then passes through a contraction passage with a smooth transition from circular cross section to rectangular shape at which point it enters the converging-diverging nozzle. After the cascade, the flow then enters the condenser where it dissipates heat to the cooling water and the outlet flow from the condenser then mixed with superheated

Table 2 The desired conditions at different points of the system for cascade tests

Description	Expected range
Mass flow rate	0.7–1.5 kg/s
Cascade inlet pressure	7–15 bara
Cascade inlet temperature	70–110 °C
Vapour quality at the condenser outlet	70%-100%
Minimum pressure at compressor suction	1.5–2 bara
Maximum compressor discharge temperature	115 °C

vapour from cascade outlet in the mixing chamber. After the mixing chamber, there are two ways for further expansion and pressure drop flow. The first one is two parallel valves that are used in combination to adjust the flow or directly flow through the screw expander. In either case, after expansion the flow passes through an orifice plate to estimate the mass-flow rate, before the flow is returned to the suction side of the compressor.

Control: The operation of the test rig is controlled by several control elements. The displacement of the compressor is adjusted by its rotational speed which is controlled by the operator via a frequency inverter. The rate of heat dissipation by the condenser is related to the cooling water flow rate which is controlled by the water valve. The valve combination after the condenser also affects the refrigerant flow rate through the system. Finally, the screw expander is directly coupled with an eddy-current dynamometer that can be controlled by setting fixed speed or fixed torque. In either case, the dynamometer speed determines the expander speed which ultimately controls the pressure ratio across the expander for a given flow rate. The combination of these control parameters (i.e., compressor speed, valves at the condenser discharge and the expander speed) will result in determining the circulating mass flow rate and the stagnation pressure and temperature upstream of the cascade. The objective of the cascade tests has been discussed earlier. Those tests are expected to achieve the desired conditions given in Table 2.

3.2 *Pre-experimental Runs*

The pre-experimental runs were conducted to evaluate the performance and operation of the rig, check the measurement equipment, and troubleshoot any unexpected problems during operation. Pressure tests were done with nitrogen with pressure up to 30 bar as a first step before charging the system with refrigerant to avoid any hazards and to ensure the functionality of the equipment.

Test description and conditions: An initial steady-state test was performed where the flow passed through three main components: compressor, cascade, and condenser. Once the rig was started and warmed, all discharge flow from the compressor was

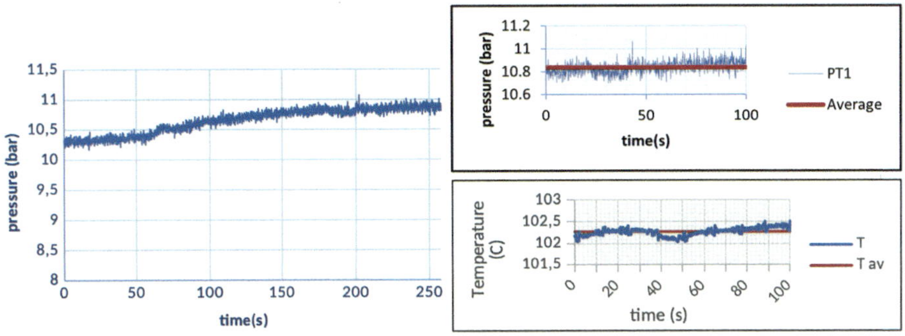

Fig. 4 Data processing example including the pressure rise during the process, selected P and T range, and average

routed through the cascade to pass through the nozzle. After this, the flow is partially condensed and then mixed with vapor extracted from the cascade discharge in the mixing chamber before returning to the compressor. During this test, the compressor was run at a constant speed of 3,500 rpm. The change in the inlet set points of the cascade was controlled by manually changing the condenser pressure using a pressure regulator. Seven different inlet set points were adjusted starting from 7 bar to 13 bar with 1 bar increments. The test duration was approximately 8–10 min for the low-pressure cases and 5 min for high pressure cases to prevent compressor overheating which could be a problem in achieving steady state and could also appear in the future tests. The pressure data was processed by averaging the last 100 s of each test to achieve stable measurements. Figure 4 shows an example for data processing where the average pressure is 10.835 bar with expanded uncertainty of ± 0.00338 bar. Regarding the temperature distribution, its planned to add thermocouples to compare their signals to the RTD signal to investigate actual oscillation and give a better indication for temperature measurement uncertainty. While efforts were made to maintain a constant total pressure, there was no control over the temperature. During this set of experiments, there is no insulation applied to the cascade, so the expansion process is considered not adiabatic.

4 Results

Experimental results: High pressure was successfully reached and the expansion in the nozzle was tested. The expansion pressure distribution along the diverging part of the nozzle starting from the throat for the seven cases is reported in Fig. 5. Calibration for the cascade pressure transducers was completed before the test, and the offset errors (zero error) have been corrected in LabVIEW by recalibration. However, there was a high thermal drift for the fourth pressure transducer which affected its reading and therefore these results have been excluded from this analysis. Figure 5 shows an

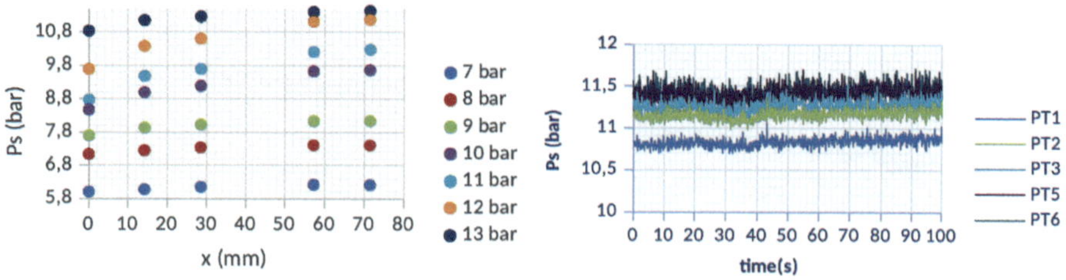

Fig. 5 Preliminary results for the initial tests. Left: pressure distribution along the nozzle for the seven tests. Right: pressure vs time for all transducers test 13 bar

Table 3 Test conditions

Test Name	Total pressure (bar)	Total temperature (°C)	Total Z
7 bar	7	76.6	0.842
8 bar	8	83.0	0.829
9 bar	9	87.8	0.815
10 bar	10	92.9	0.802
11 bar	11	96.1	0.786
12 bar	12	99.5	0.77
13 bar	13	102.3	0.754

increase in the static pressure in the diverging part of the nozzle for all tests, which indicates that the flow within the cascade section remains subsonic. A summary of the total conditions and compressibility factors from the initial test is shown in Table 3 and the location of the inlet total conditions on the T-S diagram is shown in Fig. 6. As there is no insulation applied to the cascade and the small degree of superheating is approximately between 1 and 3 °C as shown in Table 3, condensation could occur beside the walls of the cascade.

A quasi-one-dimensional isentropic expansion model (1-D) was developed to extend the understanding of the experimental results. The 1-D isentropic modelling was performed using the total pressure and temperature from the experiments, assuming the mass flow rate was calculated based on the throat conditions. The 1-D model assumed steady, frictionless, adiabatic conditions where the conservation of mass, momentum, and energy equations were applied for the converging-diverging nozzle. The equations were solved using MATLAB and the refrigerant thermodynamic properties were calculated using REFPROP version 10.0 [10]. First, the static pressures from the experimental results were used alongside the nozzle dimensions at the throat and the outlet of the nozzle to estimate the mass flow rate assuming an isentropic expansion process. The resulting mass flow rate was then compared with the isentropic choked mass-flow rate. It was found that the value calculated based on the measured outlet pressure was larger than the choked value which indicates a non-isentropic process associated with a high-pressure drop in the diverging part of

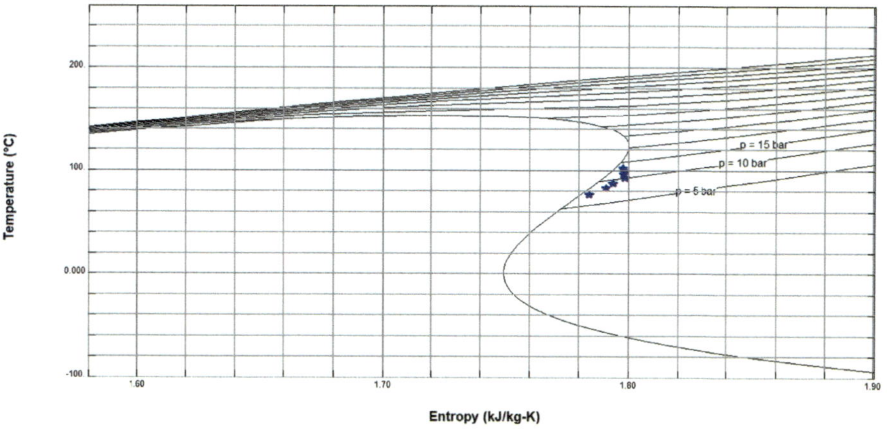

Fig. 6 T s diagram showing the location of inlet conditions

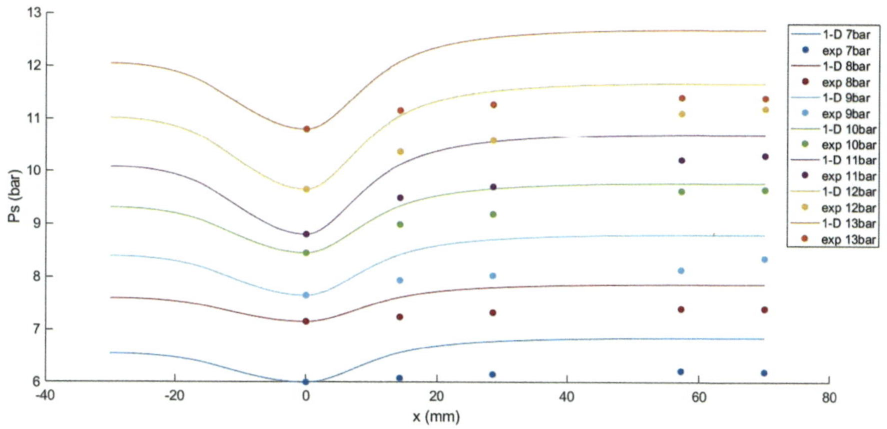

Fig. 7 Estimated pressure profiles assuming 1D isentropic flow with mass-flow rate set to the value calculated based on throat static pressure

the nozzle. On the other hand, it was found that mass-flow rate calculated based on the measured throat pressure was lower than choked mass flow rate.

It was found that the pressure profiles do not match the experiment profiles as shown in Fig. 7. There are two factors that may cause this, the mass-flow rate calculated based on the throat may not be entirely accurate due to possible boundary layer blockage and upstream pressure drops while there probable pressure drops in the diverging section that could be due to heat loss or flow separation.

5 Conclusions and Future Work

This paper provides an overview of the NextORC test facility, including its components, the different operational modes and procedures, and preliminary results. Initial commissioning of the NextORC test facility with working fluid R245fa has been successfully performed and troubleshooting for any operational problems such as leakage, pressure drops, heating and cooling, and checking the performance of different components has been completed. The initial results for the cascade performance at high pressure during commissioning experiments were reported in this paper, and the static pressure measurements obtained indicate subsonic flow conditions have been achieved.

For the upcoming steps, additional test runs will be conducted, and comparisons between the results of repeated experiments will be performed to ensure the consistency and reliability of the conducted measurements. This repetition helps also in identifying the systematic trends and detecting errors. Uncertainty analysis will also be performed, and further thermodynamic conditions will be tested to reach supersonic conditions. In order to achieve expansion process close to the isentropic expansion, insulation will be applied to the piping and the cascade to achieve adiabatic conditions. Further modelling of the test rig is also planned, including a dynamic model to study the dynamic performance of the whole test rig, alongside one-dimensional non-isentropic and CFD modelling to enable a comparison between experimental and numerical results.

References

1. Wieland, C., Schifflechner, C., Braimakis, K., Kaufmann, F., Dawo, F., Karellas, S., Besagni, G., Markides, C.N.: Innovations for organic rankine cycle power systems: Current trends and future perspectives. Appl. Therm. Eng. **225**, 120201 (2023)
2. Guardone, A., Colonna, P., Pini, M., Spinelli, A.: Nonideal compressible fluid dynamics of dense vapors and supercritical fluids. Ann. Rev. Fluid Mech. **56**, 241–269 (2024)
3. Thompson, P.A.: A fundamental derivative in gas dynamics. Phys. Fluids **14**(9), 1843–1849, 09 (1971)
4. Spinelli, A., Cammi, G., Gallarini, S., Zocca, M., Cozzi, F., Gaetani, P., Dossena, V., Guardone, A.: Experimental evidence of non-ideal compressible effects in expanding flow of a high molecular complexity vapor. Exp. Fluids **59**(8), 126 (2018)
5. Reinker, F., Hasselmann, K., aus der Wiesche, S., Kenig, E.Y.: Thermodynamics and fluid mechanics of a closed blade cascade wind tunnel for organic vapors. J. Eng. Gas Turbines Power **138**(5), 052601, 10 (2015)
6. Reinker, F., Kenig, E.Y., Passmann, M., aus der Wiesche, S.: Closed loop organic wind tunnel (clowt): Design, components and control system. Energy Proc. **129**, 200–207 (2017). 4th International Seminar on ORC Power Systems September 13-15th 2017POLITECNICO DI MILANOBOVISA CAMPUSMILANO, ITALY
7. Beltrame, F., Head, A.J., De Servi, C., Pini, M., Schrijer, F., Colonna, P.: First experiments and commissioning of the orchid nozzle test section. In: Pini, M., De Servi, C., Spinelli, A., di Mare, F., Guardone, A. (eds.) Proceedings of the 3rd International Seminar on Non-Ideal

Compressible Fluid Dynamics for Propulsion and Power, pp. 169–178. Springer International Publishing, Cham (2021)

8. Development and Commissioning of a Blowdown Facility for Dense Gas Vapours, volume Volume 3: Coal, Biomass, Hydrogen, and Alternative Fuels; Cycle Innovations; Electric Power; Industrial and Cogeneration; Organic Rankine Cycle Power Systems of Turbo Expo: Power for Land, Sea, and Air, 06 (2019)
9. Design of a Closed-Loop Optical-Access Supersonic Test Facility for Organic Vapours, volume Volume 3: Coal, Biomass, and Alternative Fuels; Cycle Innovations; Electric Power; Industrial and Cogeneration; Organic Rankine Cycle Power Systems of Turbo Expo: Power for Land, Sea, and Air, 06 (2018)
10. Lemmon, E.W., Huber, M.L., McLinden, M.O.: NIST Standard Reference Database 23: Reference Fluid Thermodynamic and Transport Properties - REFPROP. National Institute of Standards and Technology, Standard Reference Data Program, Gaithersburg, 9.0 edition (2010)

Design of Experiments to Observe Liquid-to-Vapor Expansions Exploiting Molecularly Complex Organic Fluids

Riccardo Gioia, Alessandro Romei, and Andrea Spinelli

Abstract The rising interest in novel two-phase organic Rankine cycles, promising higher efficiency in exploiting low-enthalpy heat sources compared to current single-phase cycles, faces challenges in modeling the flashing process due to insufficient experimental data. To tackle this, the TROVA blowdown wind tunnel will be modified to enable the characterization of flashing processes of molecularly complex fluids. A lumped parameter model is developed to simulate the discharge process of a representative organic fluid, namely hexamethyldisiloxane (MM), and obtain insights into the characteristic times and dimensions compatible with the facility constraints. A preliminary design for a planar converging-diverging nozzle is simulated using computational fluid dynamics, employing a homogeneous equilibrium model. Calculations demonstrate achievable complete vaporization without flow detachment for a useful test duration, corroborating the feasibility of the proposed experimental campaign.

Keywords Liquid-to-dry expansion · Flashing flows · Organic fluids · Design of experiments

1 Introduction

The necessity to halt climate change through the decarbonization of primary energy consumption has increased interest in the energy efficiency of the industrial sector, which in the EU represents 26% of primary energy use and is characterized by high losses in the form of waste heat [1]. Nearly half of this rejected heat is low grade, with

R. Gioia (✉) · A. Romei · A. Spinelli
Politecnico di Milano, Milano, Italy
e-mail: riccardo.gioia@polimi.it

A. Romei
e-mail: alessandro.romei@polimi.it

A. Spinelli
e-mail: andrea.spinelli@polimi.it

A. Giauque et al. (eds.), *Proceedings of the 5th International Seminar on Non-Ideal Compressible Fluid Dynamics for Propulsion and Power*, ERCOFTAC Series 32,
https://doi.org/10.1007/978-3-031-92695-2_19

temperatures below 300–350 °C, which can be efficiently exploited using different technologies. One of these is represented by converting waste thermal heat into electric energy employing an organic Rankine cycle (ORC).

A possible further improvement to this mature technology is substituting a single-phase expansion with a two-phase one. Following this idea, the trilateral flashing cycle [2], where the expansion starts from saturated liquid, and the partial evaporative ORC [3], where the expansion begins in the two-phase state, have been proposed. In general, the ability to control the end of evaporation guarantees a better match between the temperatures of the fluid and the heat source, reducing exergy losses and increasing cycle efficiency. Nonetheless, these improvements imply the existence of an expander capable of processing a two-phase flow with an efficiency comparable to that for a single-phase flow [4]. This requirement forces the use of volumetric expanders, thereby limiting the power output below the MW scale.

A possible way to achieve a single-stage two-phase turbine with acceptable efficiency is to leverage the dry characteristic of the saturation curve of molecularly complex fluids, which enables their complete vaporization through an adiabatic expansion in the stator [5], enabling dry operation in the rotor. However, the non-equilibrium effects that characterize this type of expansion must be considered to design a nozzle cascade with acceptable efficiency. This can be achieved by developing an experimentally based model, in analogy with successful attempts for water [6], if experimental data for organic fluids are available. Currently, such data are scarce, except for the experimental campaign carried out by Zhu and Elbel [7, 8]. The expansions studied in the aforementioned campaign begin at a low temperature and end within the saturation dome, due to the shape of the expanded fluid saturation curve, namely R134a. This contrasts with the previously mentioned need to exploit the retrograde shape of the saturation curve to achieve complete vaporization.

To fill this gap, one of the tasks of the POWHER project is to provide reliable experimental data for flashing fluids other than water at thermodynamic conditions relevant to ORC. The selection of relevant thermodynamic conditions is essential because the presence of non-equilibrium is highly dependent on the closeness to the thermodynamic critical point, making it challenging to extrapolate experimental data to different operating conditions. To obtain the necessary data and demonstrate the feasibility of complete vaporization through an adiabatic expansion, the TROVA facility at Politecnico di Milano will be purposely updated. This work highlights the main modifications to the facility and the envisaged expansion processes that will be tested in the upcoming experimental campaign.

2 Trova Modifications for Two-Phase Experiments

The TROVA plant is a blowdown wind tunnel designed to characterize the expansion of organic vapor flows [9]. As shown in Fig. 1, the test rig consists of a high-pressure vessel (HPV) where the fluid is heated and pressurized through an isochoric transformation, a test section where the fluid is expanded, and a low-pressure vessel (LPV)

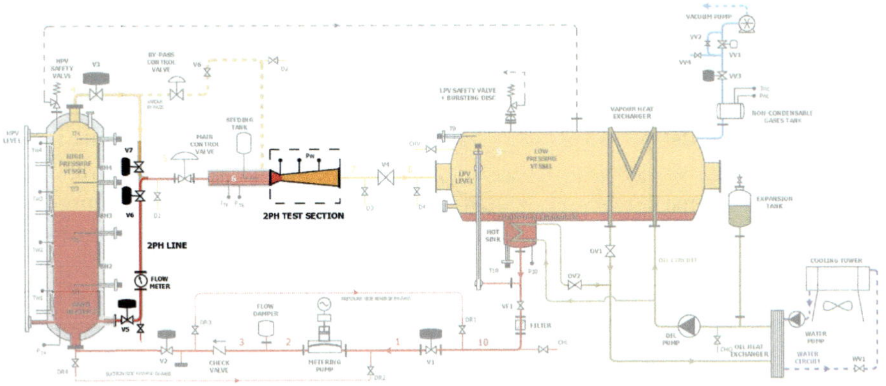

Fig. 1 Schematic of the TROVA test rig with the main modifications highlighted

where the fluid is collected and condensed at ambient temperature, ensuring a huge pressure ratio between the two vessels.

Although the current arrangement has been proven successful for the characterization of single-phase organic vapor flow [10, 11], two main modifications are necessary to observe a liquid-to-dry expansion: (i) to extract pressurized liquid from the HPV, a new pipeline connecting the bottom of the HPV to the existing main valve is necessary, (ii) a redesign of the test section, consisting in a planar converging-diverging nozzle. Similarly to the current arrangement, the nozzle will be constructed with two removable inserts to enhance the modularity of the test section.

To accurately identify the intake conditions for validation and repeatability purposes, the adiabatic expansions under study will start from a subcooled liquid state. A possible liquid-to-vapor expansion is shown in Fig. 2, considering hexamethyldisiloxane (MM) as the working fluid. Due to budget constraints, expansions starting within the saturation dome are excluded, as they would have required detailed characterizations of the initial two-phase flows in terms of vapor quality and flow regime (e.g., annular, bubbly, stratified, etc.).

3 Lumped Parameter Model of the Discharge Process and Nozzle Sizing

To gain a first insight into suitable boundary conditions to achieve complete vaporization and estimates of the nozzle gross dimensions, a lumped parameter model has been developed and detailed in the next sections.

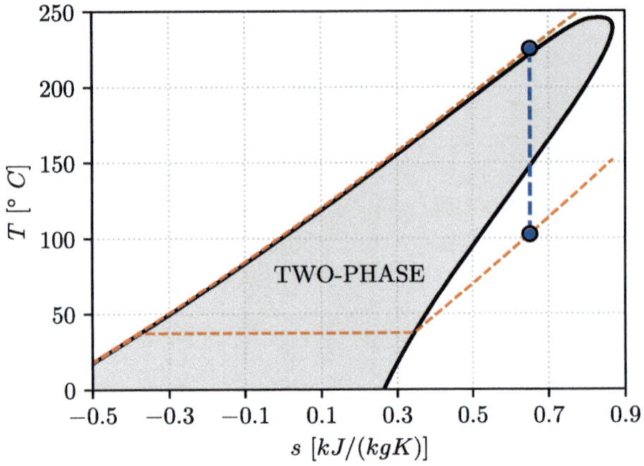

Fig. 2 Example of an isentropic liquid-to-dry expansion in the temperature-specific entropy (T, s) diagram for MM. Inlet conditions are $T = 225\,°\mathrm{C}$ and P $= 30$ bar, while outlet pressure is P $=$ 0.1 bar

3.1 Discharge Model with No Losses

The discharge process is modeled by solving continuity and energy equations using a forward time-marching method:

$$\frac{dm_{HPV}}{dt} = -\dot{m}_{out} \implies m_{HPV}^{(i+1)} = m_{HPV}^{(i)} - \dot{m}_{out}^{(i)}\Delta t \tag{1}$$

$$\frac{dU_{HPV}}{dt} = -\dot{m}_{out}h_{out}^{t} \implies u_{HPV}^{(i+1)} = u_{HPV}^{(i)} - \frac{\dot{m}_{out}^{(i)}\mathcal{H}\left(\rho_{HPV}^{(i)}, u_{HPV}^{(i)}\right)}{m_{HPV}^{(i)}}\Delta t \tag{2}$$

where (i) represents the time discretization and $\mathcal{H}$ is the function, evaluated using CoolProp [12], that correlates the specific enthalpy with the HPV thermodynamic state. The selection of (ρ, u) as an independent thermodynamic pair enables the complete definition of the flow thermodynamic state even in the presence of a two-phase flow. For a given A_{th}, $\dot{m}_{out}$ is computed by maximizing the mass flux ρV under the assumption of isentropic and adiabatic flow. To account for the increase in $\dot{m}_{out}$ due to potential non-equilibrium, which delays the formation of vapor resulting in a higher density (closer to that of a liquid), a corrective factor κ is employed, with a value ranging between 1 (complete equilibrium) and 1.5 (non-equilibrium).

The model also estimates $A_{out,n}$ necessary to reach the prescribed $P_{out,n}$, which is set to one-fourth of the saturated vapor pressure ($P_{s,v}$) obtained through an isentropic expansion from the initial total conditions. Selecting $P_{out,n}$ lower than $P_{s,v}$ ensures a margin for complete vaporization in the presence of non-equilibrium effects.

3.2 Discharge Model with Losses

To account for the effects of the pressure drop in the line connecting the HPV with the test section, a lumped parameter loss model is developed. As a result, the thermodynamic state that appears in the system of equations (1)–(2) differs from the one in the HPV. The line upstream of the test section is divided into different elements corresponding to each hydraulic device. The pressure loss of the element j is computed using Eq. (3), where the value of the loss coefficient $\xi^{(j)}$ is determined according to Idelchik [13] and $V^{(j)}$ is the absolute reference velocity for the element j. After each element, ρ is updated to consider the decrease in P due to P_{loss} and the increase in internal energy due to viscous loss. Thus, $\rho^{(j-1)}$ in Eq. (3) represents the ρ computed after the previous element.

$$P_{loss}^{(j)} = \frac{1}{2}\xi^{(j)}\rho^{(j-1)}V^{(j)2} \tag{3}$$

In the model, the distributed losses along the pipes are considered together with the local losses related to the entrance, diffusers, elbows, valves, Coriolis flow meter, porous breakwater, disk breakwater, and convergent nozzle. Moreover:

- The local pressure drop inside the ball valves, i.e. V5 and V6 in Fig. 1, is neglected because they are fully open in steady conditions, meaning that a ball valve is comparable to a straight pipe from the point of view of the loss model. The manufacturer has confirmed the validity of this assumption.
- The $P_{loss,\ CFM}$, which is the loss associated with the Coriolis flow meter (in Sect. 5 the presence of the flow meter is discussed in more detail) is treated as a parabolic function of $\dot{m}$. The analytical function is determined by interpolating the data provided by the manufacturer.

3.3 Analysis of the Discharge Process Simulation

The conditions in the HPV at t = 0 s are T_{HPV} =225 °C and P_{HPV} = 30 bar. The fluid is hexamethyldisiloxane (MM). The time window for the discharge process ranges between 0 and 20 s. The equations (1)–(2) are discretized with a time step of $\Delta t = 0.1$ s. Both inviscid and viscous cases are considered. The calculations are based on the assumption of an equivalent throat diameter of $D_{th} = 8$ mm. The worst-case scenario with $\kappa = 1.5$ is discussed, leading to a faster discharge process because of non-equilibrium effects.

The temporal evolution of P_{HPV} and $P_{s,l}$ is shown in Fig. 3a. Note that $P_{s,l} = \mathcal{P}\left(T_{in,n}, X = 0\right)$, where $\mathcal{P}$ is the function, evaluated using CoolProp, that correlates the saturated liquid pressure with the temperature at the nozzle inlet. The difference between these two pressures represents the subcooling margin from the saturated conditions. The acquisition starts at t = 5 s (black line) to exclude the initial transient

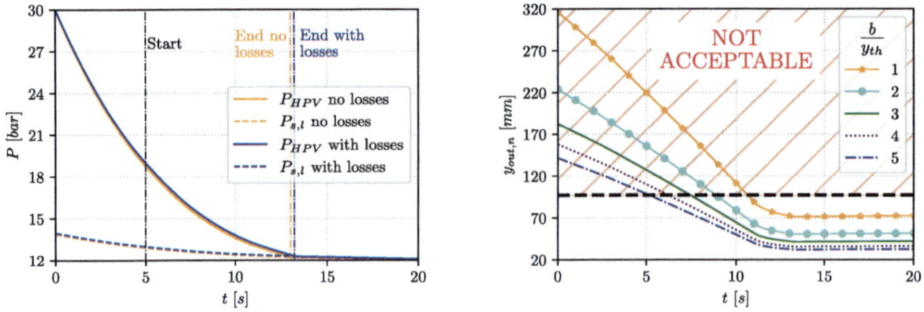

(a) (b)

Fig. 3 **a** Variation of P_{HPV} and $P_{s,l}$ at the nozzle inlet during the studied time window. **b** Variation of $y_{out,n}$ during the studied time window. The dotted black line in **b** indicates the maximum allowable $y_{out,n}$, limited by constructive constraints

effects of the fast valve opening, while the end time is reached when the phase change onset in the upstream pipeline (i.e., two-phase flows at the nozzle inlet). As P_{HPV} decreases during the emptying process, the HPV tends towards a vapor-liquid equilibrium. When this condition is reached, the subcooling is only ensured by the head of the liquid. This pressure margin from saturation is insufficient to preserve the subcooled liquid condition at the nozzle inlet. In addition, P_{HPV} and $\dot{m}$ exhibit a clear relation: the decrease of P_{HPV} is mainly due to the emptying of the HPV with a rate proportional to $\dot{m}$. At the same time, as P_{HPV} reduces, $\dot{m}$ also decreases, reaching a minimum and near-constant value when vapor-liquid equilibrium is established in the plenum. Under this light, the first derivative of P_{HPV} decreases until it flattens when flashing occurs at the nozzle inlet.

The end time for the inviscid and viscous cases is almost the same (13.0 s vs. 13.2 s), guaranteeing an acquisition length of about 8 s. At $t = 0$ s, the presence of a pressure drop related to the effect of viscous forces reduces both $P_{in,n}$ and $\dot{m}$, thus slowing down the HPV emptying compared to the inviscid one, as demonstrated by the higher value of the P_{HPV} in Fig. 3a. Nonetheless, the variations are marginal, and the effects of losses can be regarded as negligible on the test duration.

The corresponding values of $y_{out,n}$ for the viscous case at various b/y_{th} are reported in Fig. 3b. The initial idea is to design a nozzle with the value of $y_{out,n}$ computed at $t = 5$ s, ensuring that a pressure lower than the prescribed $P_{out,n}$ would always be reached during the data acquisition. Nevertheless, this is not possible, except for the geometry with $b/y_{th} = 5$, because of a constructive constraint that limits the maximum $y_{out,n}$ to the value of the internal diameter of an ASME schedule 80 pipe with a diameter of 4" (indicated by the horizontal dashed black line in Fig. 3b). However, $b/y_{th} = 5$ is not acceptable because the obtained y_{th} would be too small ($y_{th} = 3.1$ mm). In an effort to increase the throat dimension, $b/y_{th} = 3$ is selected, corresponding to a nozzle designed with the proper $y_{out,n}$ for $t = 7.5$ s and featuring an y_{th}, equal to 4.1 mm.

4 CFD Analysis

Two-dimensional and steady CFD simulations with homogeneous equilibrium model (HEM) are performed in Ansys-Fluent® to understand the flow field in a converging-diverging nozzle designed following the previous section. The thermodynamic properties are provided through look-up tables, using as primitive state variables (P, h). A more detailed explanation of the flow model used can be found in Romei et al. [14]. In this context, the use of HEM is justified by the selected expansion starting point, which is close to the critical point, where the non-equilibrium effects are expected to be negligible. Further analyses with more refined non-equilibrium models are already underway.

The simulated nozzle is designed using two Bézier curves with four control points each, one curve for the convergent part and the other for the divergent part. The passage areas at the throat and outlet are computed solving the continuity equation under the assumption of isentropic adiabatic flow conditions. Zero slopes are imposed at all two sections. The control points for the diverging sections are positioned to ensure a gradual decrease in slope going toward the outlet. Lastly, due to a constructive constraint on the test section maximum length, the total nozzle length is equal to 188 mm. A longer diverging part, with a length of 136 mm, is selected by sacrificing the converging one because of the large area ratio between the outlet section and the throat, equal to 21.7.

The nozzle is designed with the values P_{HPV}, T_{HPV}, and $P_{out,n}$ computed at $t = 7$ s with the lumped parameter model without losses. The same assumptions of Sect. 3.3 are used, except for $\kappa = 1$ to comply with the equilibrium assumption of the employed flow model.

To better explain why different values of κ have been employed in the lumped parameter model and the nozzle design for the CFD simulations, a more in-depth explanation of the effects related to the variation of κ is necessary. As explained by White [5], the dramatic decrease in the speed of sound related to the phase change causes supersonic flow in the vicinity of the geometrical throat, meaning that the actual $\dot{m}$ is equal to the choked one independently of non-equilibrium effects. However, delayed vapor formation decreases the actual X from its value in equilibrium, thereby increasing the mixture density and, correspondingly, the choked $\dot{m}$.

The nozzle geometry is simulated with three different inlet and outlet boundary conditions. The idea is to observe how a nozzle designed with the total pressure and temperature computed at $t = 7$ s behaves in off-design conditions, namely $t = 5$ s and $t = 9$ s. In all three CFD simulations only half of the passage area, discretized with a structured mesh made up of about 215000 quadrilateral elements, is simulated. Turbulence is modeled using the $\kappa - \omega\ SST$ model, assuring $y^+ < 1$ along the profile. All equations are solved using second-order methods. At the inlet, the total pressure, total enthalpy, turbulence intensity, and hydraulic diameter are assigned, while only the static pressure is assigned at the outlet. Finally, symmetry and smooth-wall boundary conditions are assigned along the axis of symmetry and nozzle profile, respectively.

Table 1 Choked mass flow rate by the lumped parameter model and the CFD simulation for the three simulated cases

t [s]	$\dot{m}_{LP}$ [kg/s]	$\dot{m}_{CFD}$ [kg/s]	Relative difference [%]
5.0	1.44	1.36	5.75
7.0	1.25	1.18	5.75
9.0	1.07	1.01	5.61

The Mach contour for the $t = 7$ s case, as illustrated in Fig. 4, indicates the absence of flow detachment, which may occur due to the significant area ratio. In the same contour, two discontinuities in the Mach number can be observed: one near the geometrical throat and another inside the divergent part, corresponding to the beginning and end of the vaporization, respectively. As widely described in the literature [15], the speed of sound computed employing the HEM is characterized by the presence of discontinuities in correspondence with saturation conditions. Another consequence of the HEM is that the two-phase speed of sound at equilibrium increases along isentropic expansions, motivating the maximum M right after the throat. Weak shocks are visible in the first half of the divergent nozzle (as shown in the zoom of Fig. 4), but they can arguably be eliminated with a more refined nozzle design via shape optimization. Finally, complete vaporization is achieved at around half of the divergent, demonstrating an adequate margin to comply with potential vapor formation delay.

Figure 5 reports the non-dimensional pressure and vapor quality trends along the nozzle axis for the three operating conditions. The shocks are merely shifted by the change in the inlet total conditions without any amplification in the shock strength. Optimizing the nozzle profile to eliminate the shock under design conditions could possibly eliminate the shock train in other operating conditions as well. The primary difference in the behavior of $P/P_{in,n}$ among the three cases is located in the first half of the divergent part. This difference can be attributed to similar P_{th} values and, consequently, the need for greater expansion with higher inlet total conditions.

Furthermore, the vaporization end occurs sooner as t increases, as shown in Fig. 5b, because the total-to-static expansion ratio to reach complete vaporization reduces. However, the worst-case scenario ($t = 5$ s) still foresees a complete vaporization at $z/L = 0.58$. Such early vaporization results from the choice to design the nozzle for $P_{out,n} = 0.25\ P_{s,v}$ as a safety measure to account non-equilibrium effects.

As expected, $\dot{m}$ decreases with the decrease of the total inlet conditions, see Table 1. This comparison also demonstrates that the lumped parameter model presented in Sect. 3 can predict the actual $\dot{m}$ with acceptable accuracy, although with an overestimation of around 5.6 – 5.7 %. This overestimation can be attributed to two-dimensional effects in the vicinity of the throat.

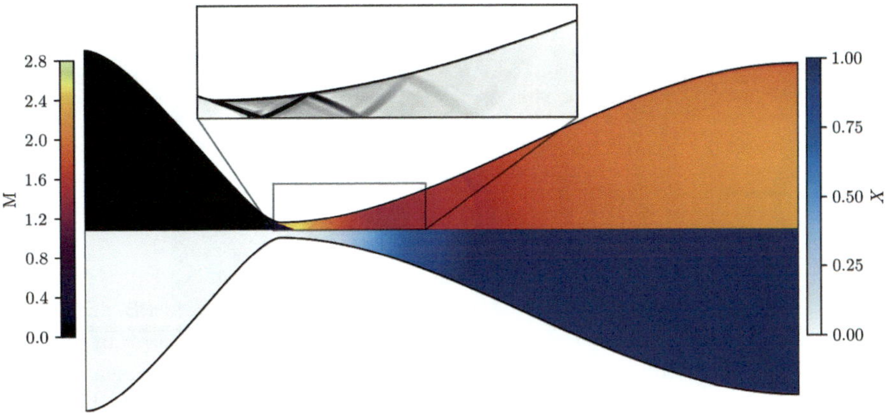

Fig. 4 M (above), X (below), and density gradient (zoom) contours for the $t = 7$ s case

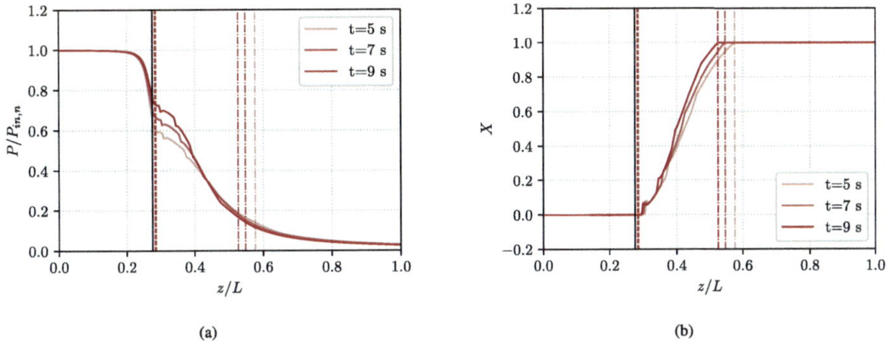

Fig. 5 **a** Variation of nondimensional P along the axis of symmetry, and **b** variation of X along the axis of symmetry. The black solid line indicates the position of the geometrical throat, the dashed line indicates the start of the phase change, while the dash-dotted line indicates the end of the vaporization

5 Research Question and Required Instrumentation

During the experimental campaign, three main research questions will be addressed as follows:

1. The demonstration of complete vaporization through adiabatic expansion will be accomplished via qualitative optical visualization. High-speed digital imaging and shadowgraphy have been selected as optical techniques due to their proven ability to visualize two-phase flows [16, 17]. Additionally, during post-processing, if the grey contrast across the two-phase interface is sufficient, the locations of incipient and complete vaporization will be determined. Two optical accesses, one on each side of the test section (Fig. 6), are necessary to perform this type of measurement.

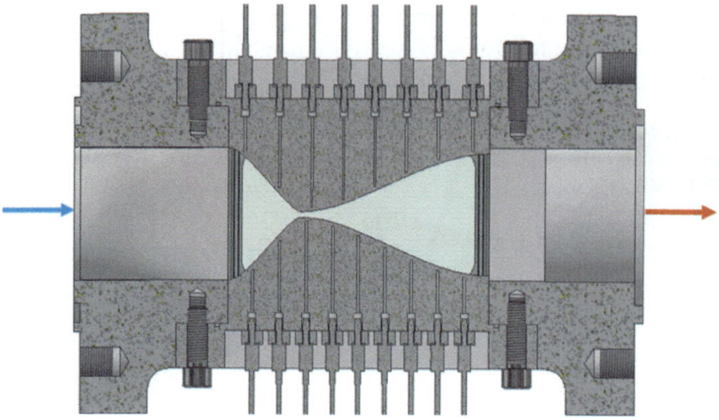

Fig. 6 Preliminary design of the new test section with the simulated profile

2. The experimental data required for the development and validation of flashing flow models will be obtained by measuring the pressure at the nozzle walls using pressure taps drilled into the inserts, see Fig. 6. These taps are asymmetrically positioned, except for two on each side, allowing for increased spatial resolution while also verifying the symmetry of the flow. Additionally, all tap positions are optically accessible, enabling the assessment of the actual flow state at each corresponding location.
3. A quantitative characterization of potential non-equilibrium between phases may be achieved by comparing $\dot{m}_{CFM}$, measured by a Coriolis flow meter mounted along the line upstream of the test section, with $\dot{m}_{HEM}$, computed numerically under the assumption of thermodynamic equilibrium and homogeneous flows. This comparison is justified by observing that the expected flow in the nozzle is supersonic and choked [5] and that delayed vapor formation is associated with a higher value of $\dot{m}$ than expected under equilibrium conditions.

Although the effectiveness of the employed techniques is widely proven in the literature, some criticalities can be identified for both the selected optical techniques and the Coriolis flow meter. Specifically:

- Regarding high-speed digital imaging and shadowgraphy, their inability to provide a complete three-dimensional reconstruction of the flow can lead to misjudgments of the actual void fraction [17]. This limitation can occur, for example, in the case of annular flow: the presence of vapor near the walls renders the flow opaque, and consequently, the liquid core cannot be detected by the camera.
- The indirect characterization of non-equilibrium through the comparison of $\dot{m}_{CFM}$ with $\dot{m}_{HEM}$ presents three conceptual problems that may possibly limit the envisaged characterization:

  1. The start of the flashing must occur before the geometrical throat to establish a meaningful connection between the choking $\dot{m}$ and the level of non-equilibrium.

2. The difference between $\dot{m}_{CFM}$ and $\dot{m}_{HEM}$ can be attributed to factors other than the presence or absence of equilibrium, such as turbulence modeling, boundary layers, and flow non-uniformity, potentially leading to an overestimation of the actual degree of non-equilibrium.
3. The characterization is valid only in the vicinity of the throat, while no information can be derived about how the non-equilibrium evolves through the expansion.

Despite the discussed limitations, the proposed experiment is considered a reasonable starting point for the characterization of organic fluid flashing flows.

6 Conclusions and Future Works

This paper outlines the design of an experimental campaign in the TROVA facility to characterize liquid-to-vapor expansions of organic fluids. The nozzle gross dimension and operating conditions are derived from modeling the discharge process with a lumped parameter approach. These conditions are then utilized to set the boundary conditions for CFD simulations of a preliminary nozzle design, employing a homogeneous equilibrium model. The CFD results demonstrate that complete vaporization without flow separation is achievable over a sufficiently extended period. These preliminary results confirm the feasibility of the planned experimental campaign and will be used to finalize the design of experiments.

The selected measurement techniques and instrumentation for characterizing the liquid-to-dry expansion of an organic dry fluid are also illustrated. The planned characterization includes assessing the actual flow state during the expansion using high-speed digital imaging and shadowgraphy, acquiring useful data for developing and validating flashing flow models by measuring the pressure at the nozzle walls, and quantitatively evaluating the degree of non-equilibrium in the vicinity of the throat by measuring the actual $\dot{m}$.

Acknowledgements We acknowledge financial support under the National Recovery and Resilience Plan (NRRP), Mission 4, Component 2, Investment 1.1, Call for tender No. 1409 published on 14.9.2022 by the Italian Ministry of University and Research (MUR), funded by the European Union NextGenerationEU Project Title: Partial evaporation ORC systems for industrial waste heat recovery - POWHER (P2022MEYKS_002) CUP D53D23018390001 - Grant Assignment Decree No. 1385 adopted on 01/09/2023 by the Italian Ministry of Ministry of University and Research (MUR).

References

1. Agathokleous, R., Bianchi, G., Panayiotou, G., Arestia, L., Argyrou, M.C., Georgiou, G.S., Tassou, S.A., Jouhara, H., Kalogirou, S.A., Florides, G.A., Christodoulides, P.: Waste heat recovery in the eu industry and proposed new technologies. Energy Proc. **161**, 489–496 (2019)
2. Elliot, D.G.: Theory and Tests of Two-phase Turbines. Jet Propulsion Laboratory Publication 81-105, Pasadena CA (1982)
3. Daniarta, S., Kolasiski, P., Imre, A.R.: Thermodynamic efficiency of trilateral flash cycle, organic rankine cycle and partially evaporated organic rankine cycle. Energy Convers. Manag. **249**, 114731 (2021)
4. White, M.T.: Cycle and turbine optimisation for an orc operating with two-phase expansion. Appl. Thermal Eng. **192**, 116852 (2021)
5. White, M.T.: Investigating the wet-to-dry expansion of organic fluids for power generation. Int. J. Heat Mass Transf. **192**, 122921 (2022)
6. De Lorenzo, M., Lafon, P., Seynhaeve, J.-M., Bartosiewicz, Y.: Benchmark of delayed equilibrium model (dem) and classic two-phase critical flow models against experimental data. Int. J. Multiphase Flow **92**, 112–130 (2017)
7. Zhu, J., Elbel, S.: Experimental investigation of a novel expansion device control mechanism: Vortex control of initially subcooled flashing r134a flow expanded through convergent-divergent nozzles. Int. J. Refrigerat. **85**, 167–183 (2018)
8. Zhu, J., Elbel, S.: Measurement of static pressure profiles of vortex flashing r134a flow expanded through convergent divergent nozzles. Int. J. Refrigerat. **108**, 258–270 (2019)
9. Spinelli, A., Pini, M., Dossena, V., Gaetani, P., Casella, F.: Design, simulation, and construction of a test rig for organic vapors. J. Eng. Gas Turbines Power **135**, 4 (2013)
10. Spinelli, A., Cammi, G., Gallarini, S., Zocca, M., Cozzi, F., Gaetani, P., Dossena, V., Guardone, A.: Experimental evidence of non-ideal compressible effects in expanding flow of a high molecular complexity vapor. Exp. Fluids **59**, 1–16 (2018)
11. Gallarini, S., Cozzi, F., Spinelli, A., Guardone, A.: Direct velocity measurements in high-temperature non-ideal vapor flows. Exp. Fluids **62**, 199 (2021)
12. Bell, I.H., Wronski, J., Quoilin, S., Lemort, V.: Pure and pseudo-pure fluid thermophysical property evaluation and the open-source thermophysical property library coolprop. Ind. Eng. Chem. Res. **53**, 2498–2508 (2014)
13. Idelchik, I.E., Ginevskii, A.S.: Handbook of Hydraulic Resistance. Begell House (2007)
14. Romei, A., Gaetani, P., Persico, G.: Computational fluid-dynamic investigation of a centrifugal compressor with inlet guide vanes for supercritical carbon dioxide power systems. Energy **255**, 124469 (2022)
15. De Lorenzo., M.: Modelling and numerical simulation of metastable two-phase flows. 5 (2018)
16. Dunn, P.F., Thomas, F.O., Davis, M.P., Dorofeeva, I.E.: Experimental characterization of aviation-fuel cavitation. Phys. Fluids **22**, 11 (2010)
17. Dash, A., Jahangir, S., Poelma, C.: Direct comparison of shadow graphy and x-ray imaging for void fraction determination. Meas. Sci. Technol. **29**, 125303 (2018)

GPSR Compliance

The European Union's (EU) General Product Safety Regulation (GPSR) is a set of rules that requires consumer products to be safe and our obligations to ensure this.

If you have any concerns about our products, you can contact us on ProductSafety@springernature.com

In case Publisher is established outside the EU, the EU authorized representative is:

Springer Nature Customer Service Center GmbH
Europaplatz 3
69115 Heidelberg, Germany

Batch number: 10755227

Printed by Printforce, the Netherlands